DENDRIMERS IN MEDICAL SCIENCE

DENDRIMERS IN MEDICAL SCIENCE

Zahoor Ahmad Parry, PhD
and
Rajesh Pandey, MD

APPLE ACADEMIC PRESS

Apple Academic Press Inc.
3333 Mistwell Crescent
Oakville, ON L6L 0A2
Canada

Apple Academic Press Inc.
9 Spinnaker Way
Waretown, NJ 08758
USA

Exclusive worldwide distribution by CRC Press, a member of Taylor & Francis Group

No claim to original U.S. Government works

Printed in the United States of America on acid-free paper

International Standard Book Number-13: 978-1-77188-441-9 (Hardcover)

International Standard Book Number-13: 978-1-315-36600-5 (CRC Press/Taylor & Francis Ebook)

International Standard Book Number-13: ISBN: 978-1-77188-442-6 (AAP Ebook)

Library and Archives Canada Cataloguing in Publication

Parry, Zahoor Ahmad, author
Dendrimers in medical science / Zahoor Ahmad Parry, PhD
and Rajesh Pandey, MD.
Includes bibliographical references and index.
Issued in print and electronic formats.
ISBN 978-1-77188-441-9 (hardcover).--ISBN 978-1-315-36600-5 (PDF)
1. Dendrimers in medicine. I. Pandey, Rajesh, 1971-, author II. Title.
R857.D36P37 2017 615.7 C2016-907921-X C2016-907922-8

Library of Congress Cataloging-in-Publication Data

Names: Parry, Zahoor Ahmad, editor. | Pandey, Rajesh, 1971- editor.
Title: Dendrimers in medical science / editors, Zahoor Ahmad Parry and Rajesh Pandey.
Description: Toronto ; New Jersey : Apple Academic Press, 2017. | Includes bibliographical references and index.
Identifiers: LCCN 2016054874 (print) | LCCN 2016055678 (ebook) | ISBN 9781771884419 (hardcover : alk. paper) | ISBN 9781315366005 (CRC Press/Taylor & Francis Ebook) | ISBN 9781771884426 (AAP Ebook) | ISBN 9781315366005 (ebook)
Subjects: | MESH: Dendrimers | Theranostic Nanomedicine--methods
Classification: LCC R857.N34 (print) | LCC R857.N34 (ebook) | NLM QT 36.5 | DDC 610.28--dc23
LC record available at https://lccn.loc.gov/2016054874

Apple Academic Press also publishes its books in a variety of electronic formats. Some content that appears in print may not be available in electronic format. For information about Apple Academic Press products, visit our website at **www.appleacademicpress.com** and the CRC Press website at **www.crcpress.com**

DEDICATION

My eyes are in tears, heart is bleeding, brain has denied thinking, and fingers are crumbling, rightly so, as we were planning for a meeting on 29th January 2016 for sharing our recent success story about the book we wrote. But who knew that we would meet in a way that I will see in you in your death bed, and my job to perform your last funeral rituals . . .

Dear Professor Rajesh Pandey, I have a unique pain at the moment, so much that I want to be way from everyone I love and just be in touch with Almighty, alone for relief. How sad and bad is that my ethics force me to write this dedication page at the moment. It hard because this should be the happiest moment of my life,with this work on this book with you . . .

Dear Pandey, I first met you on 15th November 2001, and now I have thought of you innumerable times and I tried to collect my positive and negative memories of the past 15 years of knowing you, but believe me, it always during those moments I would either get 100% or 0%. This is so because whenever I think about your positives, I get 100%, and when I tried to collect your negatives, I got 0%, because I never got a single negative. I asked the same of many people, and this opinion never changed . . .

Sir . . .

To cut long story short, at this painful movement I know that ethically I have no choice but to dedicate this book to you. Had the reasons been one or few, I would have loved to mention them, but they are so many that they will take a separate book.

With love and only love
From ZAP, family, relatives and friends.

CONTENTS

LIST OF ABBREVIATIONS

ADME	absorption, distribution, metabolism, and excretion
AMD	age-related macular degeneration
BD	Brownian dynamics
CED	convection-enhanced delivery
CEJ	cement-enamel junction
CPD	cationic phosphorus dendrimers
CTCs	circulating tumorcells
DOX	drug doxorubicin
DPPG	dipalmitoyl phosphatidylglycerol
ds-siRNA	double stranded small interfering RNA
ENM	electroactive nanostructured membranes
FA	folic acid
FI	fluorescein isothiocyanate
GI	gastro-intestinal
HCPT	hydroxycamptothecin
HSA	human serum albumin
LbL	layer-by-layer
LNA	locked nucleic acid
LPS	lipopolysaccharide
MC	Monte Carlo
MD	molecular dynamics
METH	methamphetamine
MMP-7	metalloproteinase-7
MRI	magnetic resonance imaging
MTX	methotrexate
NIR	near-infrared radiation
NMR	nuclear magnetic resonance
NO	nitric oxide
NSAIDs	non-steroidal anti-inflammatory drugs
PAA	polyacrylic acid
PAMAM	polyamidoamine

PEG	polyethylene glycol
PEG-PE	PEG-1,2-dioleoyl-sn-glycero-3-phosphoethanolamine
PG-Amine	polyglycerolamine
PGA	poly glycolic acid
PL	polylysine
PLA	poly lactic acid
PLG	poly(DL-lactide-co-glycolide)
PLL	poly-L-lysine
POPC	1-palmitoyl-2-oleoyl-sn-glycero-3-phosphocholine
PPI	poly(propylenimine)
PSS	polystyrene sulfonate
RAG	resorcylidene aminoguanidine
RDFs	radial distribution function
Rg	radius of gyration
RISC	RNA-induced silencing complex
RNAi	RNA interference
ROS	reactive oxygen species
SANS	small angle neutron scattering
SASA	solvent accessible surface area
SAXS	small angles X-ray scattering
scFv	single chain variable fragments
SEV	solvent excluded volume
SLB	supported lipid bilayers
SLNs	solid lipid nanoparticles
SNP	single nucleotide polymorphism
Vd	volume of distribution
VEGF	vascular endothelial growth factor

PREFACE

Nanomedicine, a rapidly evolving off-shoot of nanotechnology, includes the use of nanoparticles for diagnosis and therapy of a variety of diseases. Of special interest is the concept of theranostics, a promising field that combines therapeutics and diagnostics into single multifunctional formulations. This field is driven by advancements in nanoparticle systems capable of providing the necessary functionalities. Out of the many nanocarriers explored, dendrimers have proven their capabilities in local/systemic drug delivery, physical stabilization of the drug, solubility enhancement of the poorly soluble drugs, and gene delivery. Several key features of dendrimers such as tunable nanoscale formulations, excellent control over molecular structure, and availability of multiple functional groups at the periphery distinguish them as a superior choice over available polymers. The diversity of bioactives loaded in dendrimers due to covalent and non-covalent interactions, contribute to the physical forces for binding of bioactives. Further, carefully designed molecular modeling techniques have served as a fuel to foster dendrimer research. In this book, key experiments and interpretations thereof, of leaders in the field of dendrimer research from around the globe, has been compiled as a perspective with the intention of bridging the gap between basic research and applied nanomedicine. The potential role of dendrimers in combating communicable as well as non-communicable diseases is highlighted along with other pertinent examples of recent breakthroughs in dendrimer-based nanomedicine. At the same time, certain bottlenecks are also appreciated implying that direct clinical adaptations of dendrimers may be limited at present, mainly because of their high cost of production, unpredictable behavior in living organisms, unknown bioavailability, biocompatibility or pharmacokinetics, issues with therapeutic dose selection, or toxicity. Nevertheless, it is apt to say that, in the last three decades, dendrimer research has gone from infancy into adolescence, and its entry into mature adulthood awaits toiling research inputs in the coming decades.

Dr. Zahoor Ahmad Parry is thankful to the Department of Biotechnology and Council of Scientific and Industrial Research New Delhi, Gov ernment of India for financial support in the form of research grants bearing grant numbers GAP-1160, BSC-205 and MLP-1160 respectively.

ABOUT THE AUTHORS

Zahoor Ahmad Parry, PhD
Senior Scientist, Indian Institute of Integrative Medicine, Srinagar, India

Zahoor Ahmad Parry, PhD, is currently Senior Scientist at the Indian Institute of Integrative Medicine, Srinagar (Branch), in India. He has published over 25 research papers published in international journals and has written several book chapters. He has been the recipient of several research grants to study development of new drug regimens for the treatment of multi-drug resistant tuberculosis. In addition, he has participated at several international conferences and has given invited talks as well as served on several organizing committees. Dr. Parry has received several awards and honors for his research work. He is currently a member of the American Thoracic Society (ATS), the American Association of Microbiology (ASM), and the International Union against Tuberculosis and Lung Disease-NAR. His current research interests include nanotechnology-based drug delivery, drug development, and antimicrobial drug discovery.

Dr. Parry earned his PhD at PGIMER, Chandigarh, India, working on nanotechnology-based drug delivery. He also studied at the University of Kashmir and Johns Hopkins University (Baltimore, Maryland, USA).

Rajesh Pandey, MD
Professor and Head, Department of Biochemistry, MM Institute of Medical Sciences and Research, Mullana, Ambala (Haryana), India

Rajesh Pandey, MD, is currently Professor and Head, Department of Biochemistry, MM Institute of Medical Sciences and Research, Mullana, Ambala (Haryana), India. His clinical experience includes acting as House Physician for six months in the Department of Medicine, Medical College, Calcutta, India. Dr. Pandey has published two books, several book chapters, and over 100 articles in national and international journals. He has presented at a number of professional conferences and has participated in a number of workshops. He has organized webinars and meetings. He is an editorial board member of the *European Journal of Pharmaceutical and Medical Research*. Dr. Pandey earned his MD in biochemistry from the Medical College, Rohtak, India.

CHAPTER 1

INTRODUCTION: NANOTECHNOLOGY AND MEDICAL SCIENCE

ZAHOOR AHMAD PARRY, PhD, and RAJESH PANDEY, MD

CONTENTS

1.1 INTRODUCTION

Advances in knowledge of disease pathogenesis especially at the molecular level, has allowed drug designing to flourish, not only as a science but also as an art. Drugs are hardly ever administered to a patient in an unformulated state. A drug dosage formulation consists of one or more active ingredients along with other molecules termed as excipients. It has been increasingly realized that the use of excipients is as important as the drug itself. Excipients facilitate the preparation and administration, enhance the consistent release of the drug and protect it from degradation. Thus, excipients arc no longer considered to be inert substances because they can potentially influence the rate and/or extent of drug absorption and thus determine the bioavailability of the drug. The term bioavailability refers to the amount of drug available to the systemic circulation out of the total drug administered to a patient, and it is an important consideration in pharmaceutical dosage forms because the presence of the drug in the systemic circulation is essential to reach its target site and exert its therapeutic effect. A drug needs to be formulated as to extract the maximum therapeutic benefit out of it, and this is the underlying concept behind a drug delivery system. Out of the four 'Ds' in a drug delivery system, *d*rug, *d*estination, *d*isease and *d*elivery, the latter is the only variable parameter [1]. When a drug formulation is designed in such a way that the rate and/or place of drug release is altered, the formulation is called a modified release system. Alternative terms are in common usage including sustained/controlled/slow/extended/prolonged release, etc. Modified release is generally achieved by means of encapsulation. Encapsulation technology finds extensive applications in the pharmaceutical industry for the controlled release of drugs. Polymers are extensively used, both as conventional excipients and more specifically as a tool in controlled drug delivery. Polymers are broadly classified as being synthetic or natural. Poly (DL-lactide-co-glycolide) (PLG), poly lactic acid (PLA), poly glycolic acid (PGA), poly anhydrides, poly methyl acrylates, carbomer, etc., are common examples of synthetic polymers used as drug delivery vehicles. On the other hand, natural polymers include alginic acid, chitosan, gelatin, dextrins, etc. [1]. The drug release profile can be tuned depending on the choice of the drug carrier (Table 1.1). Among the most versatile polymers are PLG, alginic acid and chitosan. However, non-polymeric drug carriers, e.g., lipids in the form of solid lipid nanoparticles (SLNs) are also gaining importance. Whatever the carrier system, the ultimate aim is to improve the drug bioavailability by circumventing one or several of the possible factors known to affect the same [2] (Table 1.2).

The importance of polymeric nanoparticles and their role in the development of drug delivery systems is well established (Table 1.3). The essential difference between microparticles and nanoparticles is not merely the size but also the ability of nanoparticles to achieve a higher drug encapsulation and enhance the bioavailability of orally administered drugs. Particles with large surface area per unit mass such as nanoparticles may dissolve rapidly in the gastrointestinal tract, which can increase drug uptake, as the local concentration of drug may be higher than conventional

TABLE 1.1 Types of Polymeric Carriers Used in Drug Delivery Systems

A. Natural carriers	B. Synthetic carriers
1. Proteins and polypeptides	1. Aliphatic polyesters and hydroxy acids
• Albumin	• Polylactic acid
• Fibrinogen/fibrin	• Polyglycolic acid
• Collagen	• Poly(lactide-co-glycolide)
• Gelatin	• Polyhydroxy butyric acid
• Casein	• Polycaprolactone
2. Polysaccharides	2. Polyanhydrides
• Alginic acid	3. Polyorthoesters
• Starch	4. Polyalkylcyanoacrylate
• Dextrans/dextrin	5. Polyamino acids
• Hyaluronic acid	6. Polyacrylamides
• Chitin	7. Polyalkylcarbonates
• Chitosan	

TABLE 1.2 Variables Influencing Drug Bioavailability That Are Amenable to Improvement by Using an Appropriate Delivery System

Parameters	Examples of drugs
Low shelf-life	Ethambutol
Unpalatability	Metronidazole
Extremes of pH	Rifampicin in presence of isoniazid in acidic pH
Interaction with food	Rifampicin
Interaction with other drugs	Rifampicin
Poor solubility in intestinal fluid	Danazol
Poor intestinal absorption	Streptomycin
Extensive first pass metabolism	Propranolol
Subtherapeutic levels in plasma	Clotrimazole
Short duration of stay	Azathioprine
Distribution to non-target organs	Anticancer drugs

dosage forms. Further, nanoparticles are known to cross the intestinal permeability barrier directly via transcellular/paracellular pathways, which explain the better delivery of the encapsulated drugs into the circulation [3]. Several methods have been reported to obtain particles in the nanorange [4] (Table 1.4).

TABLE 1.3 Some Important Drugs Reported to be Incorporated in Synthetic Polymeric Nanoparticles

Drug	Category
Digitoxin	Cardiac glycoside [5]
Rolipram	Anti-inflammatory [6]
Heparin	Anticoagulant [3]
Betamethazone	Corticosteroid [7]
Enalaprilat	Antihypertensive [8]
Octyl methoxy cinnamate	Sunscreen [9]
Cyclosporine	Immunosuppressant [10]
Insulin	Hormone [11]
Praziquantel	Anti-helminthic [12]
Clotrimazole, Econazole	Antifungal [13]
Moxifloxacin	Quinolone [14]
Gentamicin	Aminoglycoside [15]
Paclitaxel	Anticancer [16]

TABLE 1.4 Various Preparation Techniques for PLG-Nanoparticles

Techniques	Merits/demerits
Emulsion/evaporation	Poor entrapment of hydrophilic drugs
Double emulsion/evaporation	Good entrapment of hydrophilic/hydrophobic drugs
Salting out	Lengthy purification process
Emulsification-diffusion	Quick process
Solvent displacement/nanoprecipitation	Poor entrapment of hydrophilic drugs
Emulsification-diffusion-evaporation	Better reproducibility of size/shape of nanoparticles

1.2 NANOTHERANOSTICS

The term theranostics refers to the integration of diagnostics and therapeutics [17]. Nanotheranostics is the application and further development of nano medicine strategies in order to achieve advanced theranostics. This implies the designing and use of various nanocarriers including micelles, liposomes, polymer conjugates, dendrimers, metal and inorganic nanoparticles, carbon nanotubes, etc. for sustained and targeted co-delivery of diagnostic and therapeutic agents. Obviously, the aim is better theranostic effects vis-a-vis fewer side effects, i.e., to diagnose and treat diseases as early as possible (when the diseases are most amenable for cure or at least treatable). Advanced theranostic nanomedicine is multifunctional, i.e., it is capable of simultaneous diagnosis and delivering drugs to the diseased cells assisted by targeting ligands and biomarkers [18]. In addition, theranostic nanomedicine may work better compared with other theranostics because of unique capabilities in a single all-in-one platform (Table 1.5) [19]. It is also emphasized that the encapsulation of one particular diagnostic/therapeutic chemical might not have the expected high therapeutic efficacy or sensitivity or specificity. Hence, multimodality nanotheranostics needs to be developed in order to make use of the merits. This can be achieved by co-encapsulation of multiple diagnostic and therapeutic modes in targeting nanomedicine platforms [20].

TABLE 1.5 Merits of theranostic nanomedicine.

- Sustained/controlled release.
- Targeted delivery.
- Better uptake by endocytosis.
- Smart delivery (stimulus responsive agent release).
- Synergetic performance (combination therapy, siRNA co-delivery, etc.).
- Quality performances (oral delivery, escape from multi drug resistance protein, autophagy inhibition, etc.).
- Multimodality diagnosis and/or therapy.

1.3 NOW AND NEXT?

The major drawbacks associated with new drug development include- an input of immense research efforts, cost and time; difficulty in targeting drug-resistant cells; and uncertainty with respect to toxicity and resistance [2, 21, 22]. This is the driving force behind the search for alternative therapeutic strategies. By employing nanotechnology, synthetic/natural carrier-based controlled release nanomedicine formulations have been developed, encapsulating key drugs against several diseases. Besides sustained release of drugs in plasma/organs, other potential advantages of the system include the possibility of selecting various routes of therapy, reduction in drug dosage/adverse effects/drug interactions, targeting drug-resistant cells, latent microorganisms, etc. On the other hand, the choice of carrier, large-scale production, stability and toxicity of the formulation are some of the major issues that merit immediate attention and resolution. Nevertheless, keeping in view the hurdles in new drug development, nanomedicine has provided a sound platform for an onslaught against communicable and non-communicable diseases. The specific role played by dendrimers will be the subsequent focus in this book.

KEYWORDS

- **bioavailability**
- **drug**
- **drug delivery systems**
- **encapsulation**
- **nanoparticles**
- **polymers**

REFERENCES

1. Pandey, R., & Khuller, G. K. (2004). Polymer based drug delivery systems for mycobacterial infections. *Current Drug Delivery, 1*(3), 195–201.
2. Ahmad, Z., & Pandey, R. (2013). Nanotechnology-based drug delivery in Combating infectious diseases: Sebastian, M., Ninan, N., Haghi, A. K. (Eds.). Recent

Advances in Nanoscience and Nanotechnology, Vol. 1: *Nanomedicine and Drug Delivery,* Apple Academic Press, Toronto, Canada. Chapter 9, pp. 97–125 (text) and 239–251 (references). ISBN: 978-1-926895-17-8.

3. Jiao, Y., Ubrich, N., Marchand-Arvier, M., Vigneron, C., Hoffman, M., Lecompte, T., & Maincent, P. (2002). *In vitro* and *in vivo* evaluation of oral heparin-loaded polymeric nanoparticles in rabbits. *Circulation, 105*(2), 230–235.
4. Bala, I., Hariharan, S., & Kumar, M. N. (2004). PLGA nanoparticles in drug delivery: the state of the art. *Critical Reviews in Therapeutic Drug Carrier Systems, 21*(5), 387–422.
5. Guzmán, M., Aberturas, M. R., Rodríguez-Puyol, M., & Molpeceres, J. (2000). Effect of nanoparticles on digitoxin uptake and pharmacologic activity in rat glomerular mesangial cell cultures. *Drug Delivery, 7*(4), 215–222.
6. Lamprecht, A., Ubrich, N., Yamamoto, H., Schäfer, U., Takeuchi, H., Maincent, P., Kawashima, Y., & Lehr, C. M. (2001). Biodegradable nanoparticles for targeted drug delivery in treatment of inflammatory bowel disease. *Journal of Pharmacology and Experimental Therapeutics, 299*(2), 775–781.
7. Horisawa, E., Hirota Kawazoe, T. S., Yamada, J., Yamamoto, H., Takeuchi, H., & Kawashima, Y. (2002). Prolonged anti-inflammatory action of DL-lactide/glycolide copolymer nanospheres containing betamethazone sodium phosphate for an intra-articular delivery system in antigen-induced arthritic rabbit. *Pharmaceutical Research, 19*(4), 403–410.
8. Ahlin, P., Kristl, J., Kristl, A., & Vrecer, F. (2002). Investigation of polymeric nanoparticles as carriers of enalaprilat for oral administration. *International Journal of Pharmaceutics, 239*(1–2), 113–120.
9. Alvarez-Román, R., Naik, A., Kalia, Y. N., Guy, R. H., & Fessi, H. (2004). Enhancement of topical delivery from biodegradable nanoparticles. *Pharmaceutical Research, 21*(10), 1818–1825.
10. Jaiswal, J., Gupta, S. K., & Kreuter, J. (2004). Preparation of biodegradable cyclosporine nanoparticles by high-pressure emulsification-solvent evaporation process. *Journal of Controlled Release, 96*(1), 169–178.
11. Bilati, U., Pasquarello, C., Corthals, G. L., Hochstrasser, D. F., Allémann, E., & Doelker, E. (2005). Matrix-assisted laser desorption/ionization time-of-flight mass spectrometry for quantitation and molecular stability assessment of insulin entrapped within PLGA nanoparticles. *Journal of Pharmaceutical Sciences, 94*(3), 688–694.
12. Mainardes, R. M., & Evangelista, R. C. (2005). PLGA nanoparticles containing praziquantel: effect of formulation variables on size distribution. *International Journal of Pharmaceutics, 290*(1–2), 137–144.
13. Pandey, R., Ahmad, Z., Sharma, S., & Khuller, G. K. (2005). Nano-encapsulation of azole antifungals: Potential applications to improve oral drug delivery. *International Journal of Pharmaceutics, 301*(1–2), 268–276.
14. Ahmad, Z., Pandey, R., Sharma, S., & Khuller, G. K. (2008). Novel chemotherapy for tuberculosis: chemotherapeutic potential of econazole- and moxifloxacin-loaded PLG nanoparticles. *International Journal of Antimicrobial Agents, 31*(2), 142–146.
15. Imbuluzqueta, E., Gamazo, C., Lana, H., Campanero, M. A., Salas, D., Gil, A. G., Elizondo, E., Ventosa, N., Veciana, J., & Blanco-Prieto, M. J. (2013). Hydrophobic gentamicin-loaded nanoparticles are effective against *Brucella melitensis* infection in mice. *Antimicrob Agents Chemother. 57*(7), 3326–3333.

16. Zhang, J., Han, J., Zhang, X., Jiang, J., Xu, M., Zhang, D., & Han, J. (2015). Polymeric nanoparticles based on chitooligosaccharide as drug carriers for co-delivery of all-trans-retinoic acid and paclitaxel. *Carbohydr Polym. 129*, 25–34.
17. Sumer, B., & Gao, J. (2008). Theranostic nanomedicine for cancer. *Nanomedicine (Lond). 3*(2), 137–140.
18. Janib, S. M., Moses, A. S., & MacKay, J. A. (2010). Imaging and drug delivery using theranostic nanoparticles. *Adv Drug Deliv Rev. 62*(11), 1052–1063.
19. Muthu, M. S., Leong, D. T., Mei, L., & Feng, S.-S. (2014). Nanotheranostics—Application and further development of nanomedicine strategies for advanced theranostics. *Theranostics 4*(6), 660–677.
20. Choi, K. Y., Liu, G., Chen, X. (2012). Theranostic nanoplatforms for simultaneous cancer imaging and therapy: current approaches and future perspectives. *Nanoscale. 4*(2), 330–342.
21. Pandey, R., & Ahmad, Z. (2011). Nanomedicine and tuberculosis: facts, flaws and future. *Nanomedicine 7*(3), 259–272.
22. Ahmad, Z., Maqbool, M., & Raja, A. F. (2011). Nanomedicine for tuberculosis: insights from animal models. *Int. J. Nano. Dim. 2*(1), 67–84.
23. Ahmad, Z., & Khuller, G. K. (2008). Alginate based sustained release drug delivery systems for tuberculosis. *Expert Opinion on Drug Delivery 5*(12), 1323–1334.

CHAPTER 2

THE CONCEPT OF DENDRIMERS

ZAHOOR AHMAD PARRY, PhD, and RAJESH PANDEY, MD

CONTENTS

21.1 INTRODUCTION

The term 'dendrimer' is an architectural motif and does not specify a particular molecule or class of compounds. The earliest reports on dendrimers date back to the pioneering publications by Fritz Vogtle [1], by Donald Tomalia [2], and by George R. Newkome [3], between 1978 to the early 1980s; during that period the term 'cascade molecules' was frequently employed for dendrimers. The salient features shared by dendrimers are given in Table 2.1.

2.2 BASIC ARCHITECTURE OF DENDRIMERS

As pointed out above, it is more appropriate to describe a dendrimer as an architectural motif. The structure begins with a central atom or group of atoms called as the core. From this central structure, the branches of other atoms ('dendrons') grow through a variety of chemical reactions. Indeed,

TABLE 2.1 Common Features of Dendrimers

- Artificial, monodisperse, compact macromolecules with multiple functional groups.
- Contain symmetric branching units built around a small molecule or a linear polymer core.
- Extensively branched molecules with a carefully tailored architecture.
- Linearly increase in diameter adopting a globular shape with increasing dendrimer generations.
- Undergo changes in size, shape, and flexibility as a function of increasing generations.
- The chemical groups in the periphery can be functionalized, thus modifying their physicochemical and biological properties.

this tree-like pattern gave the term ‘arborols’ to dendrimers. Potential sites of chemical complexation are called the ‘void space’ or ‘internal voids’ (Figures 2.1 and 2.2) [4, 5].

2.3 DENDRIMERS: OVERVIEW AND GENERAL CONSIDERATIONS

Dendrimers are a versatile class of regularly-branched macromolecules with unique structural and topologic features. Small size (typically less than 100 nm), narrow molecular weight distribution, and relative ease of incorporation of targeting ligands make them attractive candidates for drug delivery. Dendrimers have minimal polydispersity and high functionality. Similar to polymers, they are obtained by attaching several monomeric units, but unlike the conventional polymers, they have a highly branched three-dimensional architecture. Dendrimers are characterized by the presence of three different topologic sites, i.e., a polyfunctional core, interior layers, and multivalent surface [6]. The polyfunctional core, surrounded by extensive branching, has the ability to encapsulate several chemical moieties. Ammonia and ethylene diamine are two examples of core-synthesizing materials. The core may be surrounded by several layers of highly branched repeating units, such as polyethers, porphyrins, polyamidoamines, polyphenyls, and polyamino acids. The properties of

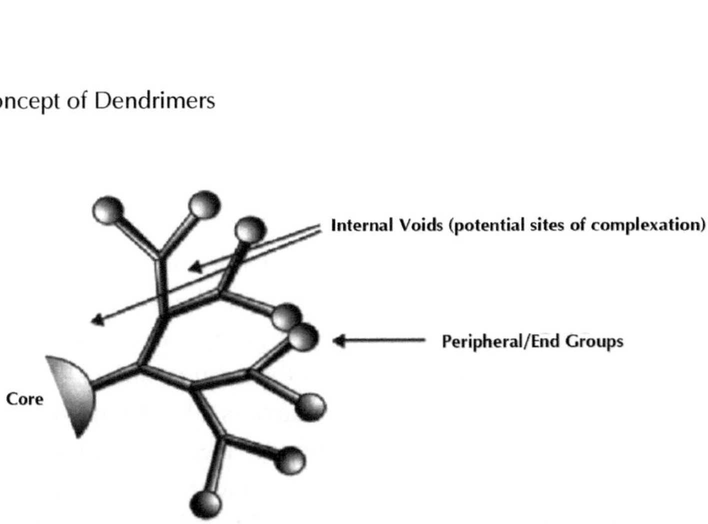

FIGURE 2.1 Basic architecture of a dendrimer. (© Abbasi et al.; licensee Springer. 2014. Reprinted with permission via the Creative Commons Attribution License. From Elham Abbasi, Sedigheh Fekri Aval, Abolfazl Akbarzadeh, Morteza Milani, Hamid Tayefi Nasrabadi, Sang Woo Joo, Younes Hanifehpour, Kazem Nejati-Koshki and Roghiyeh Pashaei-Asl. Dendrimers: synthesis, applications, and properties, Nanoscale Research Letters 2014, 9:247.)

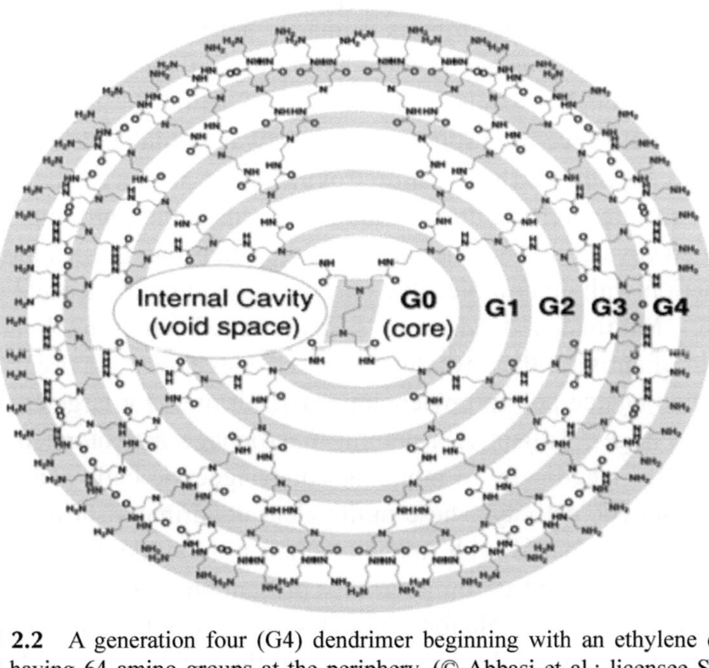

FIGURE 2.2 A generation four (G4) dendrimer beginning with an ethylene diamine core and having 64 amino groups at the periphery. (© Abbasi et al.; licensee Springer. 2014. Reprinted with permission via the Creative Commons Attribution License. From Elham Abbasi, Sedigheh Fekri Aval, Abolfazl Akbarzadeh, Morteza Milani, Hamid Tayefi Nasrabadi, Sang Woo Joo, Younes Hanifehpour, Kazem Nejati-Koshki and Roghiyeh Pashaei-Asl. Dendrimers: synthesis, applications, and properties, Nanoscale Research Letters 2014, 9:247.)

the dendrimer are predominantly based on the multivalent surface, which has several functional groups that interact with the external environment. The precise physicochemical properties of dendrimers can be controlled during synthesis by controlling the core groups, the extent of branching, and the nature and/or number of functional groups on the surface [7, 8]. As far as the use of dendrimers as carriers for antimicrobial agents is concerned, dendrimers can not only act as carriers of antiretroviral agents, but can also themselves act as antiretrovirals. Dendrimers with inherent antiretroviral activity can be synthesized by incorporating certain functional groups on their surface that can interfere with the binding of the virus to the cell. A diverse array of dendrimers with various patterns of biologic activity can be synthesized by making subtle changes, such as the type of initiator, branching unit type, dendrimer generation, linker, and surfaces [9]. The inherent antiviral activity of dendrimers has been demonstrated against influenza virus, respiratory syncytial virus, and HIV *in vitro* [10, 12]. These dendrimers primarily act by blocking viral fusion to target cells and thus act as entry inhibitors in the early stages of viral infection, although secondary mechanisms of action at later stages of the viral life cycle have been reported [6, 10]. Water-soluble dendrimers can be used as efficient carriers of antiretroviral agents which can be entrapped in the dendrimer architecture. The antiretrovirals or their prodrugs can also be grafted covalently onto the surface of the dendrimers, either alone or in conjunction with other molecules, such as targeting moieties and fluorescent tags. Multivalent dendrimeric systems have been of much interest in the field of antiviral therapy. Dutta et al. [13]developed efavirenz-loaded, tuftsin-conjugated poly(propyleneimine) dendrimers for targeted delivery to macrophages. Tuftsin is a natural macrophage activator tetrapeptide which binds specifically to mononuclear phagocytic cells and enhances their phagocytic activity. These multivalent dendrimers showed reduced cytotoxicity compared with nonconjugated poly (propyleneimine) dendrimers *in vitro*. The free amino groups present on poly (propyleneimine) dendrimers are responsible for their cytotoxicity. Conjugation of tuftsin to poly (propyleneimine) dendrimers reduced the cytotoxicity of these dendrimers, possibly by shielding the positive charges and thus preventing their interaction with cell membranes. The tuftsin-conjugated poly (propyleneimine) dendrimers showed enhanced cellular uptake by mononuclear phagocytic cells and greater anti-HIV activity *in vitro*. The same research group loaded lamivudine into mannose-capped poly(propyleneimine) dendrimers and observed a significant increase in antiretroviral activity, cellular uptake, and reduced cytotoxicity [14]. The mannose conjugation enabled the targeted delivery of the lamivudine-loaded dendrimers to macrophages containing lectin receptors on their surface. Dendrimers have been used to transfect silencing RNA (siRNA) to reduce HIV infection *in vitro*. Amino-terminated carbosilane dendrimers were used to protect and transfer siRNA to lymphocytes *in vitro* [15]. These dendrimers bind to siRNA via electrostatic interactions and protect it from RNase degradation. The V3 loop region of viral gp120 interacts with glycolipids, such as galactosylceramide, on the host cell for cell attachment and subsequent cell entry [16, 17]. Anionic polymers and dendrimers through ionic interactions with the V3 loop of gp120 interfere with viral-host cell interactions [18, 19]. One such anionic dendrimer is SPL7013, a poly-L-lysine dendrimer with naphthalene sulfonic acid terminations. It has a divalent benzhydrylamine amide of L-lysine as the core. Efficacy studies with 5% w/w SPL7013 as an aqueous gel showed that a single intravaginal dose of the formulation protected pig-tailed macaques from intravaginal simian human immunodeficiency virus infection [20].

The antiretroviral activity of sulfated oligosaccharides is very low [21]. However, sulfated oligosaccharides when attached to a dendrimer show high antiretroviral activity due to cluster effects [22]. Recently, Han et al. have developed oligosaccharide-based polylysine dendrimers with sulfated cellobiose [23]. These were shown to possess high anti-HIV activity, almost equivalent to dideoxycytidine, and low cytotoxicity.

Multivalent phosphorus-containing catanionic dendrimers with galactosylceramide analogues were developed by Blanzat et al. [24]. The influence of the multifunctional core, the alkyl chains, and the surface properties of the dendrimers on their stability, cytotoxicity, and antiretroviral properties were reported by the same group in subsequent studies [25, 26]. Galactosylceramide has a high affinity for the V3 loop of the gp120 viral envelope protein of HIV-1, and subsequently prevents viral fusion to the host cell membrane, thus acting as an entry inhibitor. Although the galactosylceramide dendrimers showed good antiretroviral activity, a low therapeutic index associated with cytotoxicity is one of the issues that need to be addressed before these can be considered promising antiretroviral agents. Reports on dendrimer-based anti-HIV therapy continue to appear [27].

Various aspects of dendrimers will be discussed in the subsequent chapters.

KEYWORDS

- **antiretroviral agents**
- **dendrimers**
- **interior layers**
- **macromolecules**
- **multivalent surface**
- **polyfunctional core**

REFERENCES

1. Buhleier, E., Wehner, W., & Vogtle, F. (1978). Cascade and nonskid-chain-like synthesis of molecular cavity topologies. *Synthesis 2*, 155–158.
2. Tomalia, D. A., Baker, H., Dewald, J., Hall, M., Kallos, G., Martin, S., Roeck, J., Ryder, J., & Smith, P. (1985). A new class of polymers: starburst-dendritic macromolecules. *Polym J (Tokyo) 17*(1), 117–132.
3. Newkome, G. R., Yao, Z.-Q, Baker, G. R., & Gupta, V. K. (1985). Cascade molecules: a new approach to micelles. *J Org Chem 50*, 2003–2004.
4. Arseneault, M., Wafer, C., & Morin, J.-F. (2015). Recent advances in click chemistry applied to dendrimer Synthesis. *Molecules 20*, 9263–9294.
5. Kannan, R. M., Nance, E., Kannan, S., & Tomalia, D. A. (2014). Emerging concepts in dendrimer-based nanomedicine: from design principles to clinical applications. *J Intern Med. 276*(6), 579–617.
6. du Toit, L. C., Pillay, V., & Choonara, Y. E. (2010). Nano-microbicides: Challenges in drug delivery, patient ethics and intellectual property in the war against HIV/AIDS. *Advanced Drug Delivery Reviews, 62*(4–5), 532–546.
7. Bosman, A. W., Janssen, H. M., & Meijer, E. W. (1999). About dendrimers: Structure, physical properties, and applications. *Chemical Reviews, 99*(7), 1665–1688.
8. Svenson, S., & Tomalia, D. A. (2005). Dendrimers in biomedical applications– reflections on the field. *Advanced Drug Delivery Reviews, 57*(15), 2106–2129.

9. Gajbhiye, V., Palanirajan, V. K., Tekade, R. K., & Jain, N. K. (2009). Dendrimers as therapeutic agents: A systematic review. *Journal of Pharmacy and Pharmacology, 61*(8), 989–1003.
10. Oka, H., Onaga, T., Koyama, T., Guo, C. T., Suzuki, Y., Esumi, Y., Hatano, K., Terunuma, D., & Matsuoka, K. (2009). Syntheses and biological evaluations of carbosilane dendrimers uniformly functionalized with sialyl [alpha](2–3) lactose moieties as inhibitors for human influenza viruses. *Bioorganic and Medicinal Chemistry, 17*(15), 5465–5475.
11. Barnard, D. L., Sidwell, R. W., Gage, T. L., Okleberry, K. M., Matthews, B., & Holan, G. (1997). Anti-respiratory syncytial virus activity of dendrimer polyanions. *Antiviral Res. 34*(2), 88–88.
12. Macri, R. V., Karlovská, J., Doncel, G. F., et al. (2009). Comparing anti-HIV, antibacterial, antifungal, micellar, and cytotoxic properties of tricarboxylato dendritic amphiphiles. *Bioorg Med Chem. 17*(8), 3162–3168.
13. Dutta, T., Garg, M., & Jain, N. K. (2008). Targeting of efavirenz loaded tuftsin conjugated poly(propyleneimine) dendrimers to HIV infected macrophages in vitro. *Eur J Pharm Sci. 34*(2–3), 181–189.
14. Dutta, T., & Jain, N. K. (2007). Targeting potential and anti-HIV activity of lamivudine loaded mannosylated poly (propyleneimine) dendrimer. *Biochim Biophys Acta. 1770*(4), 681–686.
15. Weber, N., Ortega, P., Clemente, M. I., Shcharbin, D., Bryszewska, M., de la Mata, F. J., Gómez, R., & Muñoz-Fernández, M. A. (2008). Characterization of carbosilane dendrimers as effective carriers of siRNA to HIV-infected lymphocytes. *Journal of Controlled Release, 132*(1), 55–64.
16. Harouse, J. M., Bhat, S., Spitalnik, S. L., Laughlin, M., Stefano, K., Silberberg, D. H., & Gonzalez-Scarano, F. (1991). Inhibition of entry of HIV-1 in neural cell lines by antibodies against galactosyl ceramide. *Science, 253*(5017), 320–323.
17. Harouse, J. M., Kunsch, C., Hartle, H. T., Laughlin, M. A., Hoxie, J. A., Wigdahl, B., & Gonzalez-Scarano, F. (1989). CD4-independent infection of human neural cells by human immunodeficiency virus type 1. *Journal of Virology, 63*(6), 2527–2533.
18. McCarthy, T. D., Karellas, P., Henderson, S. A., Giannis, M., O'Keefe, D. F., Heery, G., Paull, J. R., Matthews, B. R., & Holan, G. (2005). Dendrimers as drugs: Discovery and preclinical and clinical development of dendrimer-based microbicides for HIV and STI prevention. *Molecular Pharmacology, 2*(4), 312–318.
19. Moulard, M., Lortat-Jacob, H., Mondor, I., Roca, G., Wyatt, R., Sodroski, J., Zhao, L., Olson, W., Kwong, P. D., & Sattentau, Q. J. (2000). Selective interactions of polyanions with basic surfaces on human immunodeficiency virus type 1 gp120. *Journal of Virology, 74*(4), 1948–1960.
20. Jiang Y-H., Emau, P., Cairns, J. S., Flanary, L., Morton, W. R., McCarthy, T. D., & Tsai C-C. (2005). SPL7013 gel as a topical microbicide for prevention of vaginal transmission of SHIV89.6P in macaques. *AIDS Research and Human Retroviruses, 21*(3), 207–213.
21. Choi, Y., Yoshida, T., Mimura, T., et al. (1996). Synthesis of sulfated octadecyl ribooligosaccharides with potent anti-AIDS virus activity by ring-opening polymerization of a 1,4-anhydroribose derivative. *Carbohydr Res. 282*(1), 113–123.

22. Roy, R., Zanini, D., Meunier, S. J., & Romanowska, A. (1993). Solid-phase synthesis of dendritic sialoside inhibitors of influenza A virus haemagglutinin. *Journal of the Chemical Society: Chemical Communications, 24*, 1869–1872.
23. Han, S., Yoshida, D., Kanamoto, T., Nakashima, H., Uryu, T., & Yoshida, T. (2010). Sulfated oligosaccharide cluster with polylysine core scaffold as a new anti-HIV dendrimer. *Carbohydrate Polymers, 80*(4), 1111–1115.
24. Blanzat, M., Turrin, C. O., Perez, E., Rico-Lattes, I., Caminade, A. M., & Majoral, J. P. (2002). Phosphorus-containing dendrimers bearing galactosylceramide analogs: Self-assembly properties. *Chemical Communications (Cambridge), 17*, 1864–1865.
25. Blanzat, M., Turrin, C. O., Aubertin, A. M., Couturier-Vidal, C., Caminade, A. M., Majoral, J. P., Rico-Lattes, I., & Lattes, A. (2005). Dendritic catanionic assemblies: *In vitro* anti-HIV activity of phosphorus-containing dendrimers bearing galbeta1cer analogues. *Chembiochem, 6*(12), 2207–2213.
26. Pérez-Anes, A., Stefaniu, C., Moog, C., Majoral, J. P., Blanzat, M., Turrin, C-O., Caminade, A-M., & Rico-Lattes, I. (2010). Multivalent catanionic GalCer analogs derived from first generation dendrimeric phosphonic acids. *Bioorganic and Medicinal Chemistry, 18*(1), 242–248.
27. Telwatte, S., Moore, K., Johnson, A., Tyssen, D., Sterjovski, J., Aldunate, M., Gorry, P. R., Ramsland, P. A., Lewis, G. R., Paull, J. R., Sonza, S., & Tachedjian, G. (2011). Virucidal activity of the dendrimer microbicide SPL7013 against HIV-1. *Antiviral Research, 90*(3), 195–199.

CHAPTER 3

SYNTHESIS AND TYPES OF DENDRIMERS

ZAHOOR AHMAD PARRY, PhD, and RAJESH PANDEY, MD

CONTENTS

3.1 INTRODUCTION

Practically speaking, dendrimers are synthesized either by a divergent method or a convergent one [1]. In these methods, starting from a multifunctional core molecule, the dendrimer grows outward. The core molecule itself reacts with monomeric units containing a single reactive and two non-reactive groups, thus yielding the first-generation dendrimer. Subsequently, the new periphery of the growing dendrimer is activated to react with additional monomers. Thus, cascade reactions are the corner-stone of dendrimer synthesis just like in case of solid-phase peptide synthesis.

In the divergent synthesis approach employed in early days, the synthesis begins from the dendrimer core to which the arms are linked by additional building blocks in a stepwise though exhaustive manner. On the other hand, in the convergent synthesis approach, the synthesis actually begins from the exterior, i.e., with the structure that eventually would become the outermost arm in the final completed dendrimer. Therefore, according to this strategy, the final generation number is pre-determined. This necessitates the prior synthesis of a variety of branches (of requisite sizes) for each generation [2–4] (Figure 3.1).

Another interesting and feasible approach for enhancing the functionality of dendrimers is to synthesize spatially ordered dendrimer assemblies. In fact, dendrimers have been developed into assemblies such as monolayers, multilayers, vesicles, micelles, and microcapsules for constructing advanced biomaterials for various medical and non-medical applications. Further, modified and unmodified dendrimers have been employed for constructing these assemblies [3]. Among the dendrimer assemblies, the multilayered assemblies synthesized by the layer-by-layer (LbL) deposition of dendrimers have recently gained special attention due to their facile preparation and versatility with respect to their structure and

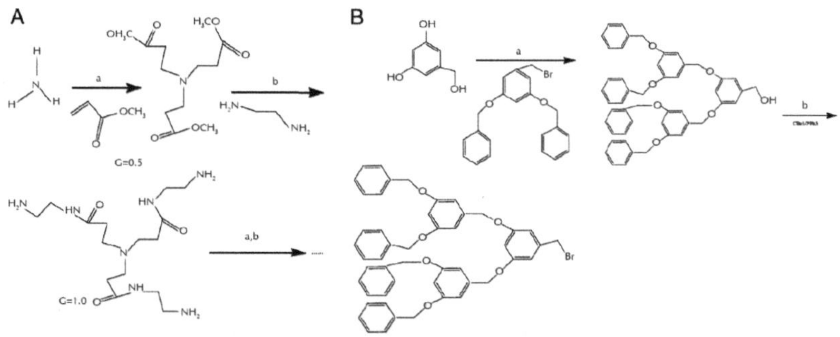

FIGURE 3.1 Approaches for the synthesis of dendrimers. (A) Divergent approach: synthesis of radially symmetric polyamidoamine (PAMAM) dendrimers using ammonia as the trivalent core; the generations are added at each synthetic cycle (two steps designated a and b), resulting in an exponential increase in the number of surface functional groups. (B) Convergent approach: synthesis of dendrons (wedges or branches) that eventually become the periphery of the dendrimer when coupled to a multivalent core in the final step of the synthesis. (© Abbasi et al.; licensee Springer. 2014. Reprinted with permission via the Creative Commons Attribution License. From Elham Abbasi, Sedigheh Fekri Aval, Abolfazl Akbarzadeh, Morteza Milani, Hamid Tayefi Nasrabadi, Sang Woo Joo, Younes Hanifehpour, Kazem Nejati-Koshki and Roghiyeh Pashaei-Asl. Dendrimers: synthesis, applications, and properties, Nanoscale Research Letters 2014, 9:247.)

function. This LbL deposition technique was developed in the early 1990s while intending to prepare multilayered nanofilms [6, 7]. Since then, the technique has flourished and LbL films have been extensively studied in materials science and technology. An advantage of LbL deposition technique is the wide selection of materials that serve as film components such as oligosaccharides/polysaccharides, DNA, proteins and synthetic polymers [8–16]. Irrespective of the surface morphology, LbL films may be conveniently deposited on the surface of a wide range of materials. Several factors determine the thickness of LbL films (Table 3.1). Alternating deposition of cationic and anionic polymers on the surface of a solid substrate via electrostatic interactions is the basis of preparing thin films. However, other factors also contribute depending on the film components (Table 3.2).

It is worth mentioning that the chemical as well as biological activities of proteins (or other functional molecules) can be preserved in the completed LbL films thereby suggesting the potential application of these films in biotechnology [23].

TABLE 3.1 Factors Determining the Thickness of LbL Films

- Number of deposits.
- pH of the bath solution.
- Ionic strength of the bath solution.
- Concentration of materials in solution.
- Deposition time.

TABLE 3.2 Interactions Stabilizing LbL Films

- Electrostatic interactions [6, 7].
- Charge-transfer interaction [17].
- Cation-dipole interaction [18].
- Hydrogen bonding [19].
- Covalent bonding [20].
- Host-guest complexation [21].
- Biological affinity [22].

3.2 ELECTROSTATIC BONDING LBL FILMS

As mentioned above, dendrimers can be assembled into multilayer LbL films via electrostatic linkage. This is because dendrimers often bear charged surface groups, e.g., amine and carboxyl residues. Hence, LbL films are synthesized using cationic dendrimers and polyanions, or anionic dendrimers in combination with polycations (Figure 3.2a). Further, oppositely charged dendrimers can also be used for constructing dendrimer LbL films without the need for other polymers (Figure 3.2b). While preparing electrostatic bonding LbL films, one should consider that dendrimers may contain charged tertiary amine groups (interior) and primary amines (periphery) [24].

LbL films made up of dendrimer and synthetic polymers have been prepared by employing a variety of polymers, including polyacrylic acid (PAA), sulfonated polyaniline, polyglycerol, polystyrene sulfonate (PSS) and an azobenzene polymer [26–30]. Figure 3.3 depicts the chemical structures of these polymers. In these cases, PAMAM and poly (propylenimine) (PPI) dendrimers (Figure 3.4) bearing positive surface charges were assembled into LbL structures via electrostatic interactions with polyanions. An interesting observation was that the polymers depos-

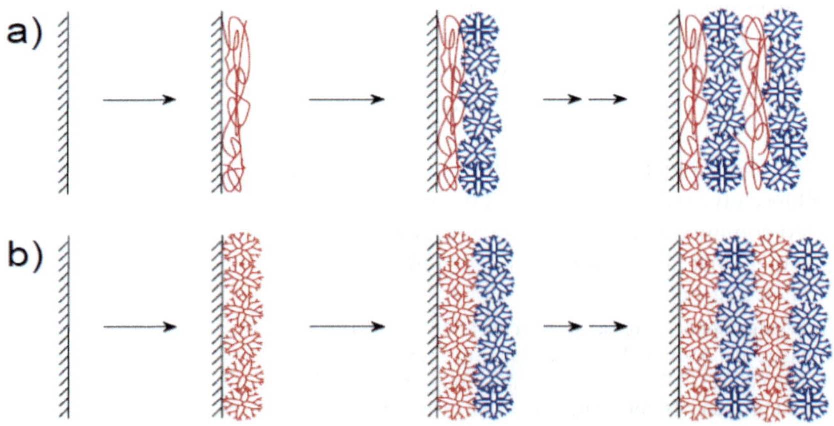

FIGURE 3.2 Preparation of (a) dendrimer/polymer and (b) dendrimer/dendrimer LbL films. (© 2013 by the authors; licensee MDPI, Basel, Switzerland. Reprinted with permission via the Creative Commons Attribution License. From Katsuhiko Sato and Jun-ichi Anzai. Dendrimers in Layer-by-Layer Assemblies: Synthesis and Applications. Molecules 2013, 18, 8440-8460; http://www.mdpi.com/1420-3049/18/7/8440/htm)

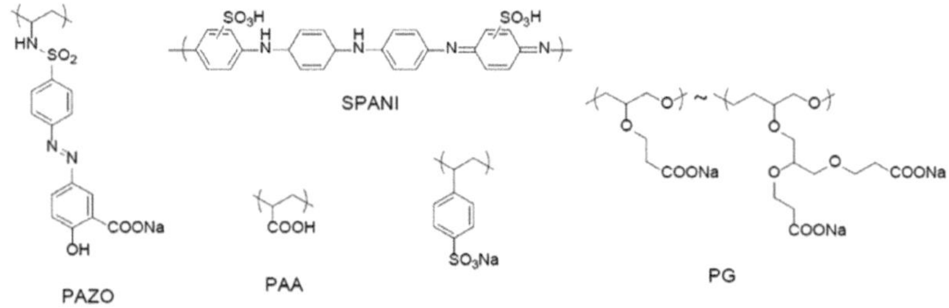

FIGURE 3.3 Structures of polyanions used for the preparation of dendrimer LbL films. (© 2013 by the authors; licensee MDPI, Basel, Switzerland. Reprinted with permission via the Creative Commons Attribution License. From Katsuhiko Sato and Jun-ichi Anzai. Dendrimers in Layer-by-Layer Assemblies: Synthesis and Applications. Molecules 2013, 18, 8440-8460; http://www.mdpi.com/1420-3049/18/7/8440/htm)

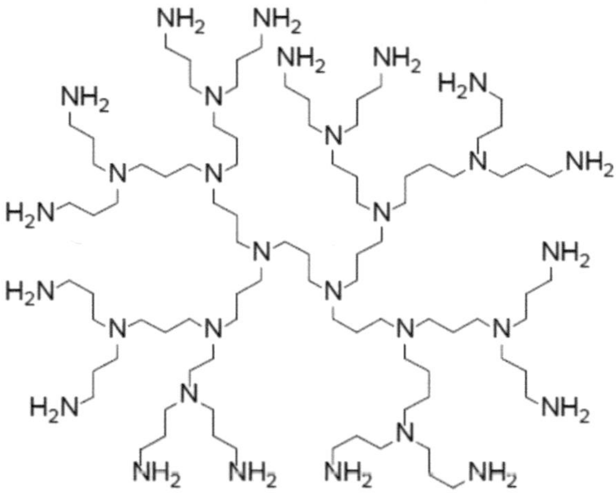

FIGURE 3.4 Structure of poly(propyleneimine) (PPI) dendrimer. (© 2013 by the authors; licensee MDPI, Basel, Switzerland. Reprinted with permission via the Creative Commons Attribution License. From Katsuhiko Sato and Jun-ichi Anzai. Dendrimers in Layer-by-Layer Assemblies: Synthesis and Applications. Molecules 2013, 18, 8440-8460; http://www.mdpi.com/1420-3049/18/7/8440/htm)

ited on the outermost surface of LbL film quite often partially desorbed when the subsequent dendrimer layer was deposited. This adsorption-desorption behavior depends on the pH as well as the ionic strength of the solutions [27–30]. Researchers have noted the pH-dependent growth of PAMAM/PAA films and obtained maximum film thickness while the film was deposited by using PAMAM solution and PAA solution at pH 8 and pH 4, respectively (at which PAMAM and PAA are partially charged). The bilayer thickness of LbL films could be tuned from 1 to 400 nm by altering the deposition pH [26]. LbL films constituting Au/Ag particle-encapsulating PAMAM dendrimers were also prepared via electrostatic bonding [31, 32].

3.3 HYDROGEN-BONDING LBL FILMS

Hydrogen bonding is yet another driving force for the assembly of LbL films [19]. Dendrimers bearing peripheral carboxylic acid residues may be used as film components since these residues can serve as H-bonding donors as well as acceptors. Researchers synthesized single-component LbL films by employing DEN-COOH (a carboxylic acid-terminated dendrimer, Figure 3.5) as the H-bonding donor/acceptor [33]. Further, two-component LbL films were synthesized by the alternate deposition of DEN-COOH (the H-bonding donor) and PVP (poly[(4-vinyl-pyridine]) as the H-bonding acceptor [34, 35]. The carboxyl-terminated PAMAM dendrimers (PAMAM-COOH, Figure 3.1) have also been used as building blocks for LbL films wherein PAMAM-COOH was

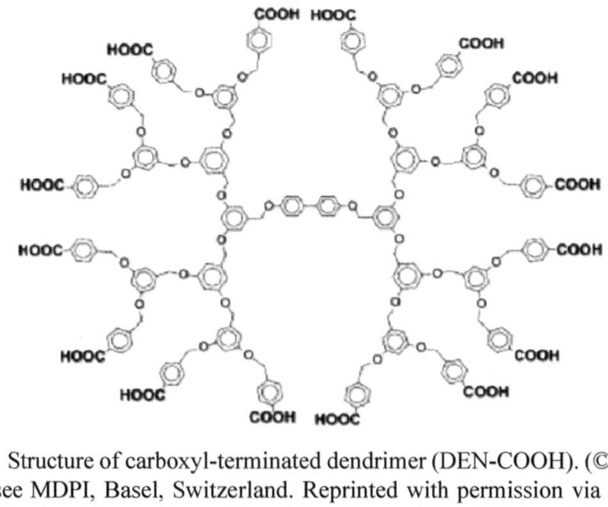

FIGURE 3.5 Structure of carboxyl-terminated dendrimer (DEN-COOH). (© 2013 by the authors; licensee MDPI, Basel, Switzerland. Reprinted with permission via the Creative Commons Attribution License. From Katsuhiko Sato and Jun-ichi Anzai. Dendrimers in Layer-by-Layer Assemblies: Synthesis and Applications. Molecules 2013, 18, 8440-8460; http://www.mdpi.com/1420-3049/18/7/8440/htm)

linked with poly carboxylic acids, e.g., PMA (poly[methacrylic acid]) and PAA. PAMAM-COOH/PMA films were synthesized at pH 4.0 via H-bonding between carboxylic acid residues in PAMAM-COOH and PMA. Nevertheless, the LbL films decomposed at pH 7 as a result of disruption of the H-bonds due to the deprotonation of the carboxylic acid residues [36, 37]. Thus, pH stability of LbL films depends on the degree of protonation of the counter polymer [38]. The researchers aptly suggested a potential application of LbL films based on PAMAM-COOH as stimulus-sensitive devices. In addition, evidence indicates that PAMAM-COOH forms monolayers/multilayers on the surface of metal substrates [39].

3.4 COVALENT-BONDING LBL FILMS

It is expected that LbL films having covalent linkages between the layers should be more stable in comparison to those prepared by electrostatic affinity and H-bonding. Covalent bonding LbL films were synthesized via the alternate deposition of PPI or PAMAM dendrimer besides poly(maleic anhydride)-co-poly(methylvinyl ether), a reactive polymer (Figure 3.6) [40, 41]. The anhydride group of the polymer reacted with the primary amines on the surface of the dendrimer forming amide and imide bonds in the LbL films. Another research group also demonstrated a protocol for preparing dendrimeric covalent-bonding LbL films based on Schiff's base formation wherein amine-terminated PAMAM along with a peroxidate ($IO4^-$)-oxidized enzyme were employed [42, 43]. A photochemical reaction was used to introduce covalent linkages in LbL films comprising of PAMAM-COOH and diazo resin [44, 45].

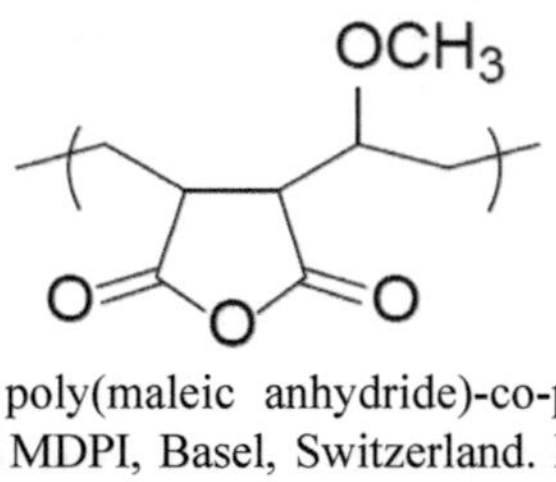

FIGURE 3.6 Structure of poly(maleic anhydride)-co-poly(methyl vinyl ether). (© 2013 by the authors; licensee MDPI, Basel, Switzerland. Reprinted with permission via the Creative Commons Attribution License. From Katsuhiko Sato and Jun-ichi Anzai. Dendrimers in Layer-by-Layer Assemblies: Synthesis and Applications. Molecules 2013, 18, 8440-8460; http://www.mdpi.com/1420-3049/18/7/8440/htm)

3.5 ALTERNATIVE APPROACHES

In another interesting report, the synthesis of dendrimer LbL films was based on coordination chemistry. A solid substrate was alternately immersed in solutions of PAMAM dendrimer and K2PtCl4 in DMSO (dimethyl sulfoxide) [46]. A linear increase in film thickness was observed up to 12–16 bilayers. Other workers established a protocol for constructing dendrimer LbL films by exploiting the biological affinity between avidin-biotin [47, 48]. Avidin is known to be a tetramer containing four identical biotin binding sites [49]. Thus, polymers labeled with multiple biotin residues may be adsorbed onto an avidin-modified surface, leaving free biotin residues for additional binding of the next avidin layer [50]. Remarkably, the alternate deposition of avidin and a biotin-labeled PAMAM dendrimer resulted in LbL films with the PAMAM dendrimer providing almost monomolecular coverage in every layer [48].

Yet another breakthrough approach emerged following the application of click chemistry for dendrimer synthesis. Click chemistry refers to the rapid and reliable generation of substances by linking small units together. Thus, it is not a single specific reaction, but outlines procedures for generating products by joining small modular units [51]. Certain desirable attributes of click chemistry are listed in Table 3.3. Thus, with a lot of enthusiasm, the click chemistry paradigm was quickly applied for synthesizing new dendritic structures. Of the various criteria defining click chemistry, high reaction enthalpy, high selectivity and atom economy combined to unravel new structures that were otherwise unattainable. The key approaches include [52]:

TABLE 3.3 Desirable Attributes of Click Chemistry

• Stereospecific, selective, modular.
• High atom economy, high chemical yields with nontoxic byproducts.
• Physiologically stable with a large thermodynamic driving force.
• Simple reaction conditions yet wide in scope.
• Employ readily available starting materials, reagents.
• No solvent used, or a nontoxic and easily removable solvent.
• Simple product isolation by non-chromatographic methods.

1. Copper-assisted azide-alkyne cycloaddition;
2. Thiol-ene and thiol-yne reactions;
3. Diels-Alder reaction.

3.6 THE CONTROVERSY OF NOMENCLATURE

There are two common nomenclatures for dendrimeric structures, viz. Newkome [53] and cascadane [54]. Both of these systems utilize the repetitive units constituting dendrimers to simplify the notation. However, with increasing dendrimer size, the notation becomes more and more complex, and difficult to interpret. Thus, additional simplification on dendrimer nomenclature, based on their repetitive topology/symmetry was proposed using a dotted cap notation [55]. The notation represents dendrimers as building blocks having a core unit with monomers and capping groups. The core is linked to the monomers forming the dendrimer framework to which the caps are attached. The dotted cap notation represents the surface of the concerned dendrimer structure by means of sequential caps (Figure 3.7).

In this case, the final notation is WZ••WZ•••WZ••WZ•••WZ••WZ •••WZ••WZ, or simply (WZ)8 in a further abbreviated notation [55].

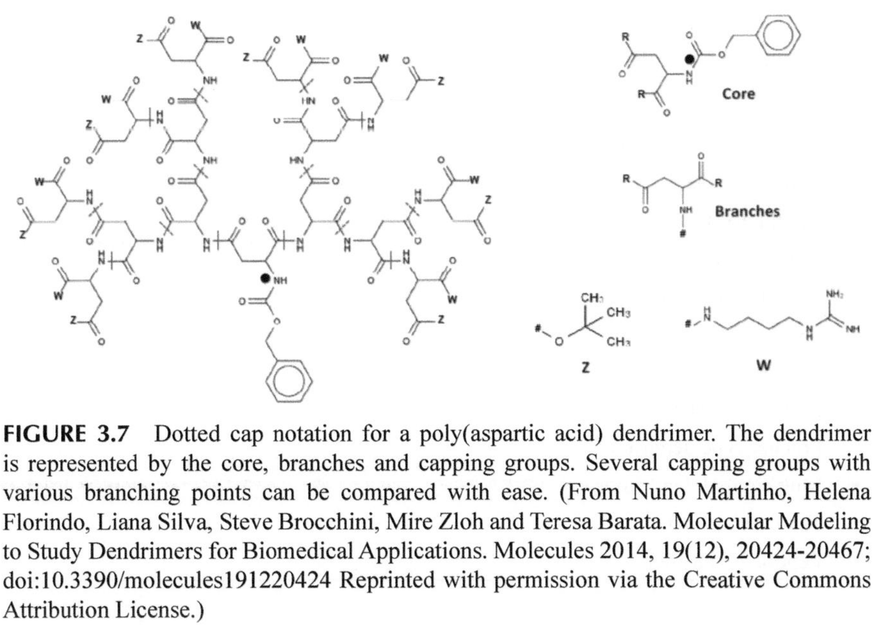

FIGURE 3.7 Dotted cap notation for a poly(aspartic acid) dendrimer. The dendrimer is represented by the core, branches and capping groups. Several capping groups with various branching points can be compared with ease. (From Nuno Martinho, Helena Florindo, Liana Silva, Steve Brocchini, Mire Zloh and Teresa Barata. Molecular Modeling to Study Dendrimers for Biomedical Applications. Molecules 2014, 19(12), 20424-20467; doi:10.3390/molecules191220424 Reprinted with permission via the Creative Commons Attribution License.)

The • represents the topological distance between the capping groups starting from the primary atom depicted in the core. However, it is clear that no information regarding the core/branching units is provided. This restricts the value of this nomenclature system for comparison purposes between similar complex structures bearing variable surface topology. Further, despite the importance of nomenclature vis a vis its description of dendrimer topology, it lacks in one important aspect, i.e., information on the 3D structure.

More popular are the terminologies based on the architectural design or specific chemical groups, often with overlapping features (Table 3.4) [56].

3.7 NOW AND NEXT?

The main reason behind the disappointingly low clinical application of dendrimers is the very synthesis of dendritic scaffolds themselves. The requirements for highly efficient reactions which are orthogonal to each other, besides the difficulties in purifying an increasingly branched molecule have hampered transfer from academia to industry. The LbL assemblies and click chemistry approaches certainly offer promise in this aspect. Nevertheless, it need not be emphasized that development of suitable molecular models would be an additional boon to synthesizing dendrimers of widespread practical utility. This aspect is covered in the next chapter.

TABLE 3.4 Common Dendrimer Terminologies

- Simple dendrimer.
- Liquid crystalline dendrimer.
- Chiral dendrimer.
- Micellar dendrimer.
- Hybrid dendrimer.
- Amphiphilic dendrimer.
- Metallo dendrimer.
- Frechet-type dendrimer.
- Multilingual dendrimer.
- Tecto dendrimer.
- Peptide dendrimer.

KEYWORDS

- **biological affinity**
- **dendrimer assemblies**
- **dendrimer core**
- **dendrons**
- **nanofilms**
- **polyamidoamine**

REFERENCES

1. Hodge, P. (1993). Polymer science branches out. *Nature 362*, 18–19.
2. Grayson, S. M., & Frechet, J. M. J. (2001). Convergent dendrons and dendrimers: from synthesis to applications. *Chem Rev 101*, 3819–3868.
3. Tomalia, D. A., Baker, H., Dewald, J. R., Hall, M., Kallos, G., Martin, S., Roeck, J., Ryder, J., & Smith, P. (1986). Dendrimers II: architecture, nanostructure and supramolecular chemistry. *Macromolecules 19*, 2466.
4. Elham Abbasi, Sedigheh Fekri Aval, Abolfazl Akbarzadeh, Morteza Milani, Hamid Tayefi Nasrabadi, Sang Woo Joo, Younes Hanifehpour, Kazem Nejati-Koshki, & Roghiyeh Pashaei-Asl. (2014). Dendrimers: synthesis, applications, and properties. *Nanoscale Research Letters 9*, 247.
5. Kano, K. (2012). Dendrimer-based bionanomaterials produced by surface modification, assembly and hybrid formation. *Polym. J. 44*, 531–540.
6. Decher, G., & Hong, J.-D. (1991). Buuildup of ultrathin multilayer films by a self-assembly process, 1 consecutive adsorption of anionic and cationic bipolar amphiphiles on charged surfaces. *Makromol. Chem. Macromol. Symp. 46*, 321–327.
7. Lvov, Y., Decher, G., & Möhwald, H. (1993). Assembly, structural characterization, and thermal behavior of layer-by-layer deposited ultrathin films of poly(vinyl sulfate) and poly(allylamine). *Langmuir 9*, 481–486.
8. Sato, K., Suzuki, I., & Anzai, J. (2003). Preparation of polyelectrolyte-layered assemblies containing cyclodextrin and their binding properties. *Langmuir 19*, 7406–7412.
9. Picart, C., Mutterer, J., Richert, L., Luo, Y., Prestwich, G. D., Schaaf, P., Voegel, J.-C., & Lavalle, P. (2002). Molecular basis for the explanation of the exponential growth of polyelectrolyte multilayers. *Proc. Natl. Acad. Sci. USA 99*, 12531–12535.
10. Crouzier, T., Boudou, T., & Picart, C. (2010). Polysaccharide-based polyelectrolyte multilayers. *Curr. Opin. Colloid Interface Sci. 15*, 417–426.
11. Sato, H., & Anzai, J. (2006). Preparation of layer-by-layer thin films composed of DNA and ferrocene-bearing poly(amine)s and their redox properties. *Biomacromolecules 7*, 2072–20767.

12. Inoue, H., & Anzai, J. (2005). Stimuli-sensitive thin films prepared by a layer-by-layer deposition of 2-iminobiotin-labeled poly(ethyleneimine) and avidin. *Langmuir 21*, 8354–8359.
13. Yoshida, K., Sato, K., & Anzai, J. (2010). Layer-by-layer polyelectrolyte films containing insulin for pH-triggered release. *J. Mater. Chem. 20*, 1546–1552.
14. Vogt, C., Mertz, D., Benmlih, K., Hemmerlé, J., Voegel, J.-C., Schaaf, P., & Lavalle, P. (2012). Layer-by-layer enzymatic platform for stretched-induced reactive release. *ACS Macro Lett. 1*, 797–801.
15. Shiratori, S. S., & Rubner, M. F. (2000). pH-Dependent thickness behavior of sequentially adsorbed layers of weak polyelectrolytes. *Macromolecules 33*, 4213–4219.
16. Liu, A., & Anzai, J. (2003). Ferrocene-containing polyelectrolyte multilayer films: Effects of electrochemically inactive surface layers on the redox properties. *Langmuir 19*, 4043–4046.
17. Shimazaki, Y., Ito, S., & Tsutsumi, N. (2000). Adsorption-induced second harmonic generation from the layer-by-layer deposited ultrathin film based on the charge-transfer interaction. *Langmuir 16*, 9478–9482.
18. Ogawa, Y., Arikawa, Y., Kida, T., & Akashi, M. (2008). Fabrication of layer-by-layer assembly films composed of poly(lactic acid) and polylysine through cation–dipole interactions. *Langmuir 24*, 8606–8609.
19. Sukhishvili, S. A., & Granick, S. (2002). Layered erasable polymer multilayers formed by hydrogen-bonded sequential self–assembly. *Macromolecules 35*, 301–310.
20. Buck, M. E., & Lynn, D. M. (2010). Free-standing and reactive thin films fabricated by covalent layer-by-layer assembly and subsequent lift–off of azlactone-containing polymer multilayers. *Langmuir 26*, 16134–16140.
21. Rydzek, G., Parat, A., Polavarapu, P., Baehr, C., Voegel, J.-C., Hemmerlé, J., Senger, B., Frisch, B., Schaaf, P., Jierry, L., et al. (2012). One-pot morphogen driven self-constructing films based on non-covalent host-guest interactions. *Soft Matt. 8*, 446–453.
22. Yao, H., & Hu, N. (2010). pH-controllable on–off bioelectroctalysis of bienzyme; layer-by-layer films assembled by cancanavalin A and glycoenzymes with an electroactive mediator. *J. Phys. Chem. B 114*, 9926–9933.
23. Tsai, H.-C., & Imae, T. (2011). Fabrication of dendrimer toward biological applications. *Progr. Mol. Biol. Trans. Sci. 104*, 101–139.
24. Niu, Y., Sun, L., & Crooks, R. M. (2003). Determination of the intrinsic proton binding constants for poly(amidoamine) dendrimers via potentiometric pH titration. *Macromolecules 36*, 5725–5731.
25. Tsukruk, V., Rinderspacher, F., & Bliznyuk, V. N. (1997). Self-assembled multilayer films from dendrimers. *Langmuir 13*, 2171–2177.
26. Kim, B. Y., Bruening, M. L. (2003). pH-dependent growth and morphology of multilayer dendrimer/poly(acrylic acid) films. *Langmuir 19*, 94–99.
27. Li, C., Mitamura, K., & Imae, T. (2003). Electrostatic layer-by-layer assembly of poly(amido amine) dendrimer/conducting sulfonated polyaniline: Structure and properties of multilayer films. *Macromolecules 36*, 9957–9965.
28. Kim, D. H., Lee, O.-J., Barriau, E., Li, X., Caminade, A.-M., Majoral, J.-P., Frey, H., & Knoll, W. (2006). Hybrid organic-inorganic nanostructures fabricated from

layer-by-layer self–assembled multilayers of hyperbranched polyglycerols and phosphorus dendrimers. *J. Nanosci. Nanotechnol. 6*, 3871–3876.
29. Khopade, A. J., & Caruso, F. (2002). Investigation of the factors influencing the formation of dendrimer/polyanion multilayer films. *Langmuir 18*, 7669–7676.
30. Casson, J. L., Wang, H.-L., Roberts, J. B., Parikh, A. N., Robinson, J. M., & Johal, M. S. (2002). Kinetics and interpenetration of ionically self-assembled dendrimer and PAZO. *J. Phys. Chem. B 106*, 1697–1702.
31. He, J.-A., Valluzzi, R., Yang, K., Dolukhanyan, T., Sung, C., Kumar, J., & Tripathy, S. K. (1999). Electrostatic multilayer deposition of a gold-dendrimer nanocomposite. *Chem. Mater. 11*, 3268–3274.
32. Esumi, K., Akiyama, S., & Yoshimura, T. (2003). Multilayer formation using oppositely charged gold- and silver- dendrimer nanocomposites. *Langmuir 19*, 76779–7681.
33. Hou, F., Xu, H., Zhang, L., Fu, Y., Wang, Z., & Zhang, X. (2003). Hydrogen-binding based multilayer assemblies by self-deposition of dendrimer. *Chem. Commun.* 874–875.
34. Zhang, H., Fu, Y., Wang, D., Wang, L., Wang, Z., & Zhang, X. (2003). Hydrogen-bonding-directed layer-by-layer assembly of dendrimer and poly(4-vinylpyridine) and micropore formation by post-base treatment. *Langmuir 19*, 8497–8502.
35. Sun, J., Wang, L., Gao, J., & Wang, Z. (2005). Control of composition in the multilayer films fabricated from mixed solutions containing two dendrimers. *J. Colloid Interface Sci. 287*, 207–212.
36. Tomita, S., Sato, K., & Anzai, J. (2008). pH-sensitive thin films composed of poly(methacrylic acid) and carboxyl-terminated dendrimer. *Sens. Lett. 6*, 250–252.
37. Tomita, S., Sato, K., & Anzai, J. (2008). Layer-by-layer assembled thin films composed of carboxyl-terminated poly(amidoamine) dendriemr as a pH-sensitive nanodevice. J. *Colloid Interface Sci. 326*, 35–40.
38. Tomita, S., Sato, K., & Anzai, J. (2009). pH-Stability of layer-by-layer thin films composed of carboxyl-terminated poly(amidoamine) dendrimer and poly(acrylic acid). *Kobunshi Ronbunshu 66*, 75–78.
39. Ito, M., & Imae, T. (2006). Self-assembled monolayer of carboxyl-terminated poly(amido amine) dendrimer. *J. Nansci. Nanotechnol. 6*, 1667–1672.
40. Liu, Y., Bruening, M. L., Bergbreiter, D. E., & Crooks, R. M. (1997). Multilayer dendrimer-polyanhydride composite films on glass, silicone, and gold wafers. *Angew. Chem. Int. Ed. 36*, 2114–2116.
41. Zhao, M., Liu, Y., Crooks, R. M., & Bergbreiter, D. E. (1999). Preparation of highly impermeable hyperbranched polymer thin-film coatings using dendrimers first as building blocks and then as in situ thermosetting agents. *J. Am. Chem. Soc. 121*, 923–930.
42. Yoon, H. C., & Kim, H. S. (2000). Multilayered assembly of dendrimers with enzymes on gold: Thickness-controlled biosensing interface. *Anal. Chem. 72*, 922–926.
43. Yoon, H. C., Hong, M. Y., & Kim, H.-S. (2000). Functionalization of a poly (amidoamine) dendrimer with ferrocenyls and its application to the construction of a reagentless enzyme electrode. *Anal. Chem. 72*, 4420–4427.
44. Zhong, H., Wang, J., Jia, X., Li, Y., Qin, Y., Chen, J., Zhao, X.-S., Cao, W., Li, M., & Wei, Y. (2001). Fabrication of covalently attached ultrathin films based on dendrimers

via H-binding attraction and subsequent UV irradiation. *Macromol. Rapid Commun. 22*, 583–586.
45. Wang, J., Jia, X., Zhong, H., Luo, Y., Zhao, X., Cao, W., & Li, M. (2002). Self-assembled multilayer films based on dendrimers with covalent interlayer linkage. *Chem. Mater. 14*, 2854–2858.
46. Watanabe. S., & Regen, S. L. (1994). Dendrimers as building blocks for multilayer construction. *J. Am. Chem. Soc. 116*, 8855–8856.
47. Anzai, L., & Nishimura, M. (1997). Layer-by-layer deposition of avidin and polymers on a solid surface to prepare thin films: significant effects of molecular geometry of the polymers on the deposition behavior. *J. Chem. Soc. Perkin Trans. 2*, 1887–1889.
48. Anzai, J., Kobayashi, Y., Nakamura, N., Nishimura, M., & Hoshi, T. (1999). Layer-by-layer construction of multilayer thin films composed of avidin and biotin-labeled poly(amine)s. *Langmuir 15*, 221–226.
49. Wilchek, M., & Bayer, E. A. (1988). The avidin-biotin complex in bioanalytical applications. *Anal. Biochem. 171*, 1–32.
50. Sato, K., Kodama, D., Naka, Y., & Anzai, J. (2006). Electrochemically induced disintegration of layer-by-layer assembled thin films composed of 2–iminobiotin–labeled poly(ethyleneimine) and avidin. *Biomacromolecules 7*, 3302–3305.
51. Kolb, H. C., Finn, M. G., & Sharpless, K. B. (2001). Click chemistry: Diverse chemical function from a few good reactions. *Angewandte Chemie International Edition 40*(11), 2004–2021.
52. Mathieu Arseneault, Caroline Wafer, & Jean-François Morin. (2015). Recent advances in click chemistry applied to dendrimer synthesis. *Molecules 20*, 9263–9294.
53. Newkome, G. R., Baker, G. R., James, K., Young, J. K., James, G., & Traynham, J. G. (1993). A systematic nomenclature for cascade polymers. *J Polymer Sci Part A: Polymer Chemistry 31*(3), 641–651.
54. Friedhofena, J. H., & Vögtle, F. (2006). Detailed nomenclature for dendritic molecules. *New J. Chem. 30*, 42–43.
55. Roberts, B. P., Scanlon, M. J., Krippner, G. Y., & Chalmers, D. K. (2008). The Dotted Cap Notation: A concise notation for describing variegated dendrimers. *New J. Chem. 32*, 1543–1554.
56. Baig, T., Nayak, J., Dwivedi, V., Singh, A., Srivastava, A., & Tripathi, P. K. (2015). A review about dendrimers: Synthesis, types, characterization and applications. *Int J Adv Pharm Biol Chem 4*(1), 44–59.

CHAPTER 4

MOLECULAR MODELING TECHNIQUES TO STUDY DENDRIMERS

ZAHOOR AHMAD PARRY, PhD, and RAJESH PANDEY, MD

CONTENTS

4.1 INTRODUCTION

Computer simulation techniques are useful to predict the properties of dendrimers at molecular level. By gaining insight into the major factors that correlate with the structure of dendrimers, one can design and optimize the properties thereof for various biomedical applications. However, it is still a great challenge to develop novel dendrimers since any modification structure is likely to modify its morphological properties and therefore its biological activity/toxicity. Hence, a better knowledge of the way these macromolecules interact with biological components is essential in order to develop safer therapeutics. The use of molecular modeling techniques can be a powerful tool to study dendrimers. However, there is a lack of dedicated software for this purpose. Here we focus on the available strategies for computational modeling of dendrimers and their applications in drug delivery.

4.2 MOLECULAR MODELING OF DENDRIMERS

In an aqueous environment, several parameters influence the architecture of dendrimers (Table 4.1), and several techniques are available to study dendrimer structure (Table 4.2) [1]. These techniques provide valuable information regarding the size as well as the molecular constituents. However, the determination of spatial configuration/geometric characterization could be challenging [2]. Besides, flexible dendrimers can possess a number of permissible configurations with a rapid interconversion between them [3]. Sometimes, specific chemical groups are used to probe their local environment, e.g., amide protons in poly(L-glutamic acid) dendrimers showed separate NMR chemical shifts and these were exploited to study their exposure to solvent (in two different generations) by altering the temperature [4], besides obtaining information about flexibility/association of lipidic peptide dendrons (G3) [5].

Computational techniques have the potential to provide valuable insight into the complex properties of dendrimers. Even by using theoretical models or experimental data, simulations can clarify mechanisms of biological interactions at molecular level. A major advantage of molecular modeling is that it allows the investigator to control every important parameter (e.g., ions, pH, structure of dendrimer) that is involved in their biological activity [6–9]. This opens a promising vista for interpretation as well as validation of experimental data with respect to the design and characterization of biological interactions [10].

TABLE 4.1 Parameters Influencing Dendrimer Architecture in Aqueous Environment

• Generation number
• Spacer length
• Density groups
• Void space
• Branching units
• Charge/hydrophobicity of surface terminal groups
• Valence of counterions in solution
• pH of solution

TABLE 4.2 Techniques to Study Dendrimer Structure

• Nuclear magnetic resonance (NMR)
• Mass spectroscopy
• Infrared and Raman spectroscopy
• Fluorescence spectroscopy
• Small angle neutron scattering (SANS)

Since dendrimers can be viewed as protein-like sequence of monomers, in a 3-dimensional assembly, their structure can be generated in a way similar to that of proteins and modeling techniques, e.g., molecular docking studies, can be applied [11, 12]. Docking studies shed more light on the interaction of dendrimers with drugs and ligands. Docking of many hydrophobic molecules (resveratrol, curcumin, genistein) showed that the interaction was generally in the hydrophobic portion of the dendrimer with limited hydrophilic interactions through the hydroxyl groups [13]. Docking studies contribute to enhance dendrimer design, permitting the selection of potential groups that may increase the affinity [14, 15]. Partially glycosylated PAMAM dendrimers were found to dock with MD-2 protein in a way that it was possible to prevent its interaction with lipopolysaccharide (LPS) [16].

The critical step while using molecular simulations techniques is to be sure about the main goals for the intended study. With the current access to computational skills, it is possible to carry out more and more complex simulations, in a reasonable time scale [17]. Prediction of dendrimer properties can be done under varying ensemble configurations (temperature, pressure, volume) that can often be challenging because of the high density/conformations that need to be defined. It goes without saying that depending on the time-scale as well as the properties to be evaluated, there is no single method that would provide all the necessary information. Dendrimers can be simulated either alone or in a defined concentration in the solvent, in the presence of small drugs/proteins/nucleic acids/lipid bilayer. It requires different scales of the number of atoms to simulate. Thus, the computational cost may differ. Ideally, these simulations should be carried with all the atoms taken into consideration (all-atom simulations) in order to get finer details of the interactions involved (such as hydrogen bonds). Coarse-grained simulations are a useful tool to study the interactions with lipid bilayers [18, 19].

To provide a general modeling technique to study dendrimers, there is a need for a suitable nomenclature that could accurately describe their topology and structure. To begin with, the naming and description of the 2-dimensional structure of these molecules is still not accepted universally. IUPAC nomenclature is an acceptable way of naming molecules and enables the exact description of almost any type of molecule. Though this nomenclature could be applied to dendrimers, it gets less clear with the increase of dendrimers' size. Further, it does not capture all the required structural features (e.g., distribution of residues). Nodal nomenclature (based on graph-theory) is able to describe dendrimers but is not widely used [20].

In principle, any chemistry-based drawing software would be sufficient to construct a depiction of the dendrimer. Unfortunately, manual assembly of these hitherto complex macromolecules is not only tedious but also highly prone to mistakes and understandably, the probability of errors increase with higher generation dendrimers. To manage this task, there are four main dedicated software packages to tackle the sequential assembly of molecules:

1. Gromacs [21].
2. XPLOR [22].
3. Starmaker (part of Silico) [23].
4. Dendrimer Building Toolkit (utilizing AMBERTOOLS) [24].

Gromacs has commonly been used for the dynamic simulation of protein structure while XPLOR is employed for structure generation based on NMR or X-ray data. The last two packages are the ones dedicated to dendrimer assembly. Regardless of the type of software package chosen, the topology and the parameters of the starting monomers need to be defined for the force field that is to be employed. The entire information pertaining to each individual atom and the way they are brought together initially in the monomers and subsequently along as a molecule then needs to be elaborated. It should be emphasized that the topology and parameter files are specific for each force field, and these files may limit the ability to interconvert between the four assembly packages.

Of late, a protocol was described that helps in dendrimer construction by describing the dendrimer as well as the link between monomers in a sequence, by using the XPLOR-NIH program (Figure 4.1) [25]. For this, the monomers are constructed in the initial stages, accomplished by common programs meant for molecular drawing. These monomers need to be defined within both the topology and parameter files as per the XPLOR format. The entire information (regarding every individual atom and the

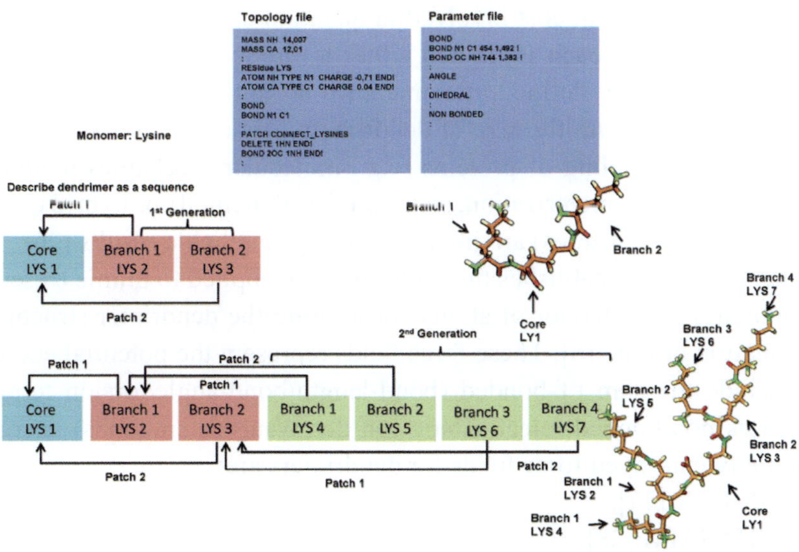

FIGURE 4.1 Constructing dendrimers of various types/generations using XPLOR. (From Barata, T. S., Brocchini, S., Teo, I., Shaunak, S., Zloh, M. (2011). From sequence to 3D structure of hyperbranched molecules: Application to surface modified PAMAM dendrimers. J. Mol. Model. 17, 2741–2749. Reprinted with permission from Springer.)

way they are brought together initially in monomers and further along as a larger molecule) is described to be used in standard molecular dynamics methods. The software then acquires the sequence and its connectivity, and starts to assemble each monomer using a simulated annealing protocol so that it can minimize potential clashes in the developing structure.

4.3 SIMULATION OF DENDRIMERS

After obtaining either the Cartesian (x, y, z) or the internal coordinates of the whole dendrimer/the monomers, the structures are minimized. Because dendrimers possess a large number of atoms, establishing the starting conditions to perform the simulations is an uphill task (due to steric overlaps as well as biased local minima). Besides, appropriate stereochemistry needs to be checked otherwise this can result in erroneous initial structures which are not corrected during the simulation process. At present, there are multiple approaches to carry out molecular simulations. They are based on quantum/molecular mechanics/molecular dynamics. But in the case of dendrimers, molecular mechanics/dynamics are commonly used owing to the high computational cost involved in quantum calculations. The quantum mechanics approach is limited either to the lowest generation dendrimers or to clearly defined monomers (in an initial conformation). To overcome hurdles with the size of dendrimers, strategies similar to those employed for modeling of proteins (e.g., molecular mechanics methods having reliable parameterization, semi-empirical methods or hybrid quantum/molecular mechanics) are likely to yield give reliable results [26].

The force field establishes the forces that are applied to mimic (resemble/simulate) the behavior of atoms constituting the dendrimer structure, as a function of time [6]. These force fields represent the potential energy and include the sum of bonded (bond-length/bond-angle/torsion terms) as well as non-bonded (electrostatic/van der Waals interactions) energy. Commonly employed force fields for dendrimers are:

1. AMBER [27].
2. CHARMM [28].
3. GROMOS [29].
4. MARTINI [30].

5. CVFF [31].
6. OPLS [32].
7. DREIDING [33].

An overview of the various force fields applied to dendrimers is given in Table 4.3. Force fields employ different protocols for parameterization and can either refer to general atom types, or to particular classes of molecules. It is, therefore, very important to be sure that the parameters suit the system, as the choice of force field may lead to different outcomes.

All-atom simulations have been carried out with PAMAM dendrimers (from G2 to G6) by using different force fields such as DREIDING/COMPASS/CVFF, and coarse-grained models (using MARTINI) [50]. To compare the performance of each force field, the radius of gyration (Rg) can be measured and compared to experimental determined values of small angles X-ray scattering (SAXS) [51]. All Rg values for different all-atom force fields are lower compared to the Rg values determined using SAXS. Almost similar values are obtained for COMPASS/CVFF force fields. However, DREIDING force field is the least reliable for small dendrimers (G2–G4) though similar to the others for higher generations (G5, G6). The scaling of size in case of DREIDING force field is similar to previous Brownian dynamic simulations. The fractal dimensions (space occupied) are also similar to those determined experimentally. For these reasons, reports suggest that DREIDING is the best for computing Rg for PAMAM dendrimers.

The utility of DREIDING force field to predict the behavior of a PAMAM generation 4 dendrimer in solution has been reported. As per the report, the radius of gyration was independent of the pH [52]. As compared to coarse-grain simulations, the Rg was reported to be similar to those obtained by DREIDING force field. These studies substantiated the usefulness of the coarse-grain model particularly when larger length scales are simulated. Last but not the least, the surrounding environment needs to be defined as precisely as possible. This includes the presence of water or some other solvent, ions, lipid membranes or even molecules such as drugs. Although a dendrimer can be simulated in vacuum, the process is not ideal due to the confounding effects of solvent polarization on its structure. In the absence of well-defined solvents and counter-ions, extra care has to be taken since mere approximations may lead to non-physiological

TABLE 4.3 Softwares Used to Build and Simulate Various Types of Dendrimers

Dendrimer type	Dendrimer construct software	Dendrimer simulation software	Force field	Reference
Conformational analysis				
PAMAM G2 to G6	Insight II	Insight II CHARMM	CVFF CHARMM	34
PAMAM G50% and 90% acetylated	Insight II	AMBER8	GAFF	35
Glycosylated PAMAM G3.5	XPLOR	Desmond	OPLS_2005	36
Triazine G3 and G5 with DOTA terminals	AMBER 11	AMBER 11	GAFF and parm99	37
Pegylated triazine dendrimers linked with paclitaxel	Material Studio 5	AMBER 11	parm99	27
Carboxylic modified PAMAM G5 with gold, fluorescein isothiocyanate (FI) and folic acid (FA)	Insight II	Insight II	CVFF	38
PAMAM G5 with methotrexate	CHARMM	CHARMM	CHARMM	39
Poly(L-lysine) and Poly(amide) G4 dendrimers	Starmaker (Silico)	NAMD	OPLS-AA	40
Dendrimer-small molecule interactions				
PAMAM G5 with different terminal groups + methoxyestradiol	Insight II	Insight II	CVFF	31
PAMAM G4 + polyphenols	ChemOffice Ultra 6.0	HyperChem Pro 7.0	MM+	13
PAMAM G3 + nicotinic acid, nicotinate and 3-pyridinium carboxylate	HyperChem	NAMD	CHARMM27	41
Peptide dendrimers + hydroxypyrene trisulfonate butyrate ester	CORINA	GROMACS	GROMOS-96 43a1	11
PAMAM G5-Folic acid + Morphine and Tramadol	ICM	NAMD	CHARMM 27; ParamChem	42

TABLE 4.3 (Continued)

Dendrimer type	Dendrimer construct software	Dendrimer simulation software	Force field	Reference
PAMAM G5 + salicylic acid, L-alanine, phenylbutazone, primidone	DBT/AMBER	AMBER9	GAFF	43
Poly(L-lysine) G4 dendrimer + doxorubicin	ChemBioOffice	Desmond	OPLS-AA	44
Dendrimer-nucleic acid interactions				
Triazine G2 dendrimers + siRNA or DNA	AMBER 10	AMBER 10	Parm99	45
PAMAM ssDNA	AMBER 7	AMBER 7	AMBER 95 (DNA) DREIDING (dendrimer)	46
PAMAM G0 and G1 + siRNA	Material Studio 5	AMBER9	Ff99 FF for RNA GAFF for dendrimers	7
PAMAM G7 + siRNA	Material Studio 5	AMBER10	GAFF (non-standard residues); parm99	47
Dendrimer-protein interactions				
Glycosylated PAMAM G3.5 + MD-2 protein	XPLOR	Desmond	OPLS_2005	16
Dendrimer-lipid bilayer interactions				
Acetylated and non-acetylated PAMAM G5 and G7 + DMPC	Insight II	GROMACS	MARTINI and adapted MARTINI	48
PAMAM G3 and G5 with different acetylation levels + DPPC	Insight II	GROMACS	MARTINI	18
PAMAM G3 with amine, acetyl and carboxyl terminals + DMPC	CHARMM	CHARMM	CHARMM27 (lipid) and CHARMM 22 (dendrimer)	49

dendrimer conformations. It has been observed with PAMAM G2 dendrimers that modifying the unbound cut-off distance and dielectric constant leads to Rg different from those systems with explicit solvent [34].

Another process to simulate the solvent is through an intermediate approximation technique of a implicit/explicit (hybrid) solvation model. This approach utilizes explicit solvent only in the layer nearer to the dendrimer [2, 53]. This approach can be used to study the structural conformation of dendrimers in a variety of solvents. The advantage is the ability to reduce the computational costs and achieving accurate solvation in the dendrimer interface [2]. After parameters are decided and set, the actual simulation is permitted to run. Although several methods have been used, the majority of simulations on dendrimers have been Brownian Dynamics (BD), Monte Carlo (MC) and Molecular Dynamics (MD) [54, 55]. Their features are given in Tables 4.4 and 4.5.

Molecular dynamics reduce computed simulation time as compared to QM by offering simplifications which assume that molecules interact as particles through classical motion mechanics. Before carrying out the actual measurement, it is advisable to go for energy minimization. This can be accomplished with a brief simulation with restrictions imposed on the degrees of freedom, e.g., in case of a peptide bond, conditions can be restrained to maintain planarity. At this juncture it is necessary to be careful otherwise the initial structure may end in a local minimum and that can be difficult to reverse [56]. Another approach that can ensure that the polymer branches are farthest away from each other can be the addition of NOE constraints during monomer assembly by using XPLOR [25]. After minimization, the suitable equations of motion are employed in iterative time-steps to simulate the dendrimer as well as the surrounding system with desirable characteristics. Depending on the model simplification, e.g., all-atom simulation, coarse-grained simulation, etc., a variety of information can be derived from the simulations. Such simulation yields relevant information like the 3D configuration and detailed atomistic interactions. Though coarse-grained simulations offer greater simplification permitting the system size as well as the simulation time-scale to be enhanced while concomitantly providing realistic details [6], they do not provide information regarding hydrogen bonding [6]. However, this method is still a valuable tool because it is valid for large systems (including lipids and

TABLE 4.4 Brownian Dynamics

- Uses simplifications allowing longer time-scale calculations.
- Individual dendrimers are treated as Brownian particles.
- Dendrimers are evaluated for friction in the surrounding solvent.
- Used to study the polyelectrolyte association between charged G3 and G4 dendrimers as well as a linear polymer having opposite charge.
- Polymer chain can be adsorbed to the dendrimer in a higher quantity than what is required to neutralize the dendrimer.

TABLE 4.5 Monte Carlo and Molecular Dynamics

- Used to study the performance of dendrimers when applied to biological systems, such as structural configurations and thermodynamic calculations.
- Dendrimers are defined with an initial configuration and subsequently minimized towards the least free energy.
- MC simulations employ random but iterative atomic displacements of the initial configuration which generates a new energy value; this value is accepted/rejected by association of a probability function by using Boltzmann statistics.
- The above function depends on whether the procedure is performed with the number of particles, volume and temperature constant, or with the number of particles, pressure and temperature constant.
- The system thus endeavors to find a minimum potential energy configuration.
- The number of degrees of freedom can be minimized to reduce the computational demand.

dendrimers). Further, as a result of decreasing the number of particles as well as the number of degrees of freedom, the time-step could be significantly increased. After a simulation is performed, there are special features that could be obtained for analysis (Table 4.6). Collectively, these features define the profile of a given dendrimer.

RDFs are useful to evaluate dendrimers as drug delivery systems because they provide additional insight into the distribution of various constituents of the system. A peak in the representation indicates the distance from the center of mass at which the atoms remain in a prolonged locked conformation. In contrast, a diffuse pattern means a homogeneous distribution all over the area of interest or a possible molecular movement. RDF is suitable to study the distribution of water molecules, ions and drugs within the interior of the dendrimers [57]. It also counts the terminal group

TABLE 4.6 Special Features Obtained Post-Simulation

- Radial distribution function (RDFs).
- Solvent accessible surface area (SASA).
- Solvent excluded volume (SEV).
- Radius of gyration (Rg).
- Shape descriptors.
- Counting the number of hydrogen pairs.
- Mechanistic interactions.
- Associated thermodynamic parameters.

distribution which is highly valuable to explore the exposure of particular groups with targeting functions [27]. Thus, the effect of peripheral groups has been studied on the distance (radial) of folic acid from the center of mass, showing that depending on the nature of the surface group, folic acid is more or less exposed to the surface [58]. It can also be used to assess the hydrodynamic radii of the solvated dendrimer [11].

SASA and SEV also yield valuable information pertaining to dendrimer structure. They allow the determination of shape, the available non-solvent but accessible internal space, and the accessibility of chemical groups of interest to the solvent. Thus, if the release of a group is sensitive to the solvent, this information can be valuable to design dendrimers that burry or expose this group. Such a case was evaluated by MD for the availability of labile linkers on various PEGylated dendrimers, to know if the linker was available, and might therefore act as a pro-drug [27]. Further, SASA and SEV examine the surface of the molecule with a spherical solvent probe and can also provide information regarding the internal cavities, besides estimating the number of molecules the dendrimer can carry [59]. This is useful to characterize various types of dendrimers and determine which one has the maximum potential. This strategy has been applied to two families of denamide and denurea dendrimers; the measurement of the internal cavity size/volume was used to determine the possible number of molecules that would accommodate in the cavities [59].

Finally, Rg permits the comparison of simulation data and the experimental data, i.e., validation of the simulation model. When values are similar, it is understood that the simulation forces are well defined. Analysis of

Rg values also provides insight into the swelling/shrinkage of the dendrimer under different conditions [52]. The radius of gyration can be computed from the gyration tensor which provides insight into dendrimer shape, e.g., spherical or elliptical [56].

4.4 MOLECULAR DOCKING OF DENDRIMERS

Before carrying out the docking of drugs into the dendrimer, it is possible to judge the potential of the dendrimer to fit the drug within. To understand this concept, two types of dendrimers viz. oxy-urea and oxy-amide groups with variable branch lengths were evaluated for the cavity sizes available for host interaction using anti-parasitic drugs [59]. Equilibrated dendrimers were made available from MD and the interior cavity dimensions were calculated from the difference between van der Waals volumes and solvent-excluded volumes. To avoid overestimating the molecular dimensions, an adequate probe size was initially determined on the little inflexion of the SASA curve versus sphere radius. Measurements of cavity size were dependent on the increase of generation but not significantly for aliphatic chain-derived branches. In addition, oxy-urea dendrimers were less porous compared to oxy-amide, possibly due to intra-molecular interactions [59]. The oxy-amide dendrimer was more suited for the incorporation of molecules because it was more flexible and possessed spacious cavities.

The determination of cavity size can also be used to rapidly estimate the maximum number of therapeutic molecules that could be incorporated into the dendrimer [12]. Docking scores can be used to calculate the free energies of individual binding sites. Thus, modified PAMAM dendrimers when blindly docked with the chemical curcumin in Autodock, using a grid box (0.3750 Å spacing), it was found that it could accept a maximum of 5 molecules and the most favorable energies could also calculated by the docking score [60]. An alternative approach can be used in which the docking conformations are initially prepared to give a starting point structure (for MD simulations). The drugs can be put in the interior of the dendrimer instead of randomly placed close to the dendrimer expecting that the drug would find the docking site in a short period of time. This is difficult to achieve for large systems owing to the density of monomers on the interior. However, this approach may prevent the possible clash of atoms which can occur when trying to manually introduce the drug into the dendrimers' voids. By using this random approach, four drugs (salicylic acid, L-alanine, phenylbutazone, primidone) were docked to the interior of PAMAM G5 dendrimer. The grid box was limited to the center in AutoDock Vina [43]. The best docking score conformations for individual drugs were used as initial structure for further MD simulations with AMBER using explicit water as solvent. Another interesting technique that has been described to incorporate drugs into dendrimers is to create cavities artificially inside the dendrimer. This can be achieved by applying force to a selected number of atoms followed by inserting the drugs into the cavities (with docking software such as AutoDock) [61].

Dendrimers may also be treated as a ligand for a protein to be considered as receptor. For example, a protein model of homodimeric A2A receptor linked to CGS21680 was used to study whether multiple copies of the agonist linked to a PAMAM G3 could simultaneously occupy both the subunits of the receptor. To make this kind of a docked molecule, a small portion of the dendrimer was initially docked into the protein and subsequently, the rest of the dendrimer was attached with the docked agonists already in place. Some clashes of overlapping branches occurred with the protein which was manually adjusted, followed by minimization and MD simulation [62].

In contrast to the above approach, full size partly glycosylated as well as non-glycosylated (serving as negative control) PAMAM dendrimers were used as ligands for docking against human MD-2 glycoprotein (the target). This was accomplished using Patchdock and Hex softwares. As both the software packages were based on rigid docking, 20 different dendrimer conformations could be obtained from MD. The partial glycosylation promoted better shape complementarity vis-a-vis showing a larger number of interactions when compared to the non-glycosylated form. Further, the docking interaction (calculated with HEX) demonstrated that the partly glycosylated dendrimer co-operatively interacted with residues located in the MD-2 entrance pocket. This revealed that not only shape but perhaps, more importantly, the electrostatic interactions were critical for the biological activity of the dendrimers. The docked structures were subsequently used as a basis for the MD simulations [16]. These examples highlight the multi-utility potential of docking methods (Table 4.7).

TABLE 4.7 The Multi-Utility Potential of Docking Methods

• Estimate free energy bindings.
• Identify binding site locations.
• Explore the therapeutic potential of a dendrimer.
• Preliminary filtering tool to optimize dendrimer structure.
• Study the possible use of a dendrimer to bind a putative guest molecule.
• Check if drugs can be incorporated into a dendrimer.
• Evaluate a range of drugs fitting the desired *in vivo* profile of a dendrimer.

4.5 MODELING DENDRIMERS FOR BIOMEDICAL APPLICATIONS

The design/development of dendrimers for biomedical applications is enhanced when there is an understanding of their properties within a physiological milieu.

4.6 IMPACT OF SOLVENT AND DENDRIMER TOPOLOGY

One of the main applications of molecular modeling for the studying dendrimers has been to better understand their conformational behavior in solution. As the expected dendrimer structure/topology is apparently dependent on the factors inherent to the dendrimer (such as dendrimer generation, monomers length, chemical properties, etc.) as well as the factors related to the solvent (type of solvent and salt, ionic strength, etc.), the combination of these factors may affect the binding pockets as well as the possible surface moieties [59–62]. There has been a major stress to understand the behavior of PAMAM dendrimers in solution. These dendrimers are commercially available and there is a huge collection of experimental data that may be used for comparison. As a result, various levels of theory and force fields have been applied to the PAMAM dendrimers, from which Rg have been measured and compared to experimental values. Although PAMAM dendrimers may have limited biomedical applications (due to their inherent toxicity at high generations) [63], such studies are nevertheless important for methodological development and validation.

Notably, PAMAM dendrimers show different protonation levels at varying pH value because they are made up of both primary and tertiary amines [64, 65]. Because representing water/ions explicitly is computationally demanding, the use of lower levels of theory may be considered and the solvent may be treated implicitly. One possible way to overcome this problem is to carry out an all-atom simulation (with explicit solvent) on a smaller generation, and then compare the Rg and atom distribution with simulations carried out with different implicit parameters [34]. From such studies, it has been found that the use of a distance-dependent dielectric constant lacking a cut-off distance has the best similarities with the explicit simulations for neutral/low pH dendrimers, and was acceptable for high pH [64]. One key issue is that the implicit treatment of water might not fully represent the degree of solvation, including the diffusion of water inside the dendrimer branches (which contribute to swelling) [62]. A good solvent system is necessary for a reliable prediction of dendrimer size/conformation as well as the assessment of solvent penetration into dendrimer void spaces [66]. Water behaves differently depending on its relative position with a dendrimer. Thus, in the case of PAMAM dendrimers, 3 classes of water have been reported: (i) buried water; (ii) surface water (at dendrimer-solvent interface); and (iii) bulk water (solvent) [66, 67]. Water is enthalpy favored near the dendrimer while buried water has lower entropy in relation to bulk water. Therefore, the binding of water molecule to dendrimers results in release of the difference of free energy [67].

Even though water molecules penetrate inside dendrimers and compete for hydrogen bonds between dendrimer residues [35, 62], other factors such as the employed force field may also contribute to the correct prediction of experimentally obtained data. Different types of solvent can promote structural alterations of dendrimers that may impact their properties. MD simulations were carried out in explicit solvent for dendrimers with the core having a linear PEG chain. Depending on the solvent (i.e., methanol/THF) the PEG could be more extended or more compacted in order to increase or decrease its interaction with the solvent. This caused the burying of the dendrimer [2]. Although SASA was not measured, the measurement of Rg revealed that in methanol, the PEG core extended outwards with a tendency to wrap around the dendrimer like that observed in snapshots acquired from the final frames of simulation. This study provided the rationale for experimental results [68] wherein two structure forms could be suggested (wrapping around and. loops to the exterior). The importance is that it gives a mechanistic view of the changes in the material in various solvents and can be employed to design and customize dendrimers responding differently in different solvent systems.

A combination of the effects of solvent and the topological features of dendrimers was reported using coarse-grained MD with different dendrimer generations (G4–G7), different spacer lengths (1–6 molecules) and charge (neutral/partly and fully charged). It was observed that neutral dendrimers had a more spherical as well as compact structure when compared to charged dendrimers (which had void spaces). Expectedly, the resulting space left is being available to encapsulate drugs. Further, with increasing size, higher volume would be obtained. Indeed, it was found that with an increase in PAMAM generation from 4 to 6, the internal volume increased [69], as also observed by measuring the solvent excluded surface of the simulated dendrimers [62]. By modifying the size of the spacer in the core, an enormous influence on the size of the dendrimer and on its internal structure was seen (changes from a compact and nearly spherical to a "blob-like" structure [70]. Predicting such a stimulus-triggered behavior has a major impact for applications as drug delivery systems because new dendrimers can be designed to encapsulate drugs under specified conditions of pH or different types of solvents, which prevent the burst release *in vivo*. It may also potentiate the design of a stimulus-responsive mechanism, e.g., pH-triggered release.

4.7 IMPACT AND VERSATILITY OF THE END GROUPS

Due to the hyper-branched topology of the dendrimers, there are many terminal groups that could be tailored to simultaneously participate in many specific or non-specific interactions. A common practice to overcome the potential toxicity of multiple charges at the dendrimer surface is to attach other groups, e.g., acetyl, PEG or lipid moieties. The end groups may also be modified to carry targeting moieties such as folic acid [58]. End group modifications may affect the overall dendrimer structure and result in altered host-guest interactions or inefficient presentation of the targeting moieties. As an example, a commonly observed phenomenon for PAMAM dendrimers is the back-folding of their end groups [62, 71, 72] as measured by the density distribution of the end groups from the center of mass. If the intended purpose is to protect a labile group from the external environment, such a structure would be effective. However, for a linked moiety (e.g., targeting molecule), most likely this structural arrangement would be ineffective. Both the cases can be elucidated by molecular modeling approaches which can exemplify the location of these groups as well as the available surface area to the solvent, besides providing insight regarding the potential of modified terminal groups to be available for interaction with a possible biological target.

With the purpose of examining the effect of end capping moieties on the exposition of folic acid to interact with the receptor, MD simulations were performed with CVFF in implicit solvent which was treated with a distance-dependent dielectric constant [58]. For the PAMAM G5 dendrimers with folic acid/terminal groups composed of amine/hydroxyl/carboxyl/acetamide, simulations revealed that both the dendrimers with charged group's internalized folic acid as measured between the mean distances of folic acid from the center of mass, compared to Rg. In contrast, in case of acetamide derivative, the surface groups were extended away thereby suggesting the potential capacity to interact with the receptor. The hydroxyl terminated dendrimer had higher exposure of the folic acid to the surface although less pronounced when compared with the acetamide derivative. Taken together, these measurements could correlate directly with cell internalization assay of the dendrimers [58]. Likewise, all-atom simulations with the CHARMM 22 force field on PAMAM G3 dendrimers terminated either with amine/hydroxyl groups, or hydroxyl groups having 4 methotrexate (MTX) molecules, were tested for the availability of the drug (MTX). The Rg for PAMAM-OH and PAMAM-MTX was less compared to PAMAM-NH_2. The snapshots of equilibrated simulations suggested that MTX remained at the surface. The results positively correlated with the binding affinity of the dendrimers determined experimentally [73].

Considering another approach, MD simulations were carried out to design double-labeled Dendrimers in which the probes did not interfere with each another (either by quenching or in any other way) [74].

A PAMAM dendrimer was functionalized with carboxy-fluorescein as well as tetramethyl-rhodamine. PEG spacers were introduced to study if it could maintain both the fluorophores at a certain distance. The simulations were performed in AMBER 11 with either no PEG at all, or 44 monomers of PEG. An explicit water model was used with sufficient ions to neutralize the dendrimer charges together with 150 mM NaCl to mimic the experimental conditions. Analysis of the radial distribution function as well as the fluorescein-rhodamine distance suggested a suitable distance only in the presence of PEG spacers. These results were in accordance with the quantum yield assessed by optical spectroscopy [74]. Based on validation data obtained from the modeling studies, these dendrimers were used to probe the physiological and pathological environments by measuring the fluorescence and assuming that both the probes did not interfere with each another. Such an approach bears good potential for developing diagnostic tools. A practical and very useful application of this kind of molecular modeling of dendrimers (bearing several types of sensing probes) is in guided surgery by luminescence of the tumor cells [75]. Dendrimers can be designed to have sensing probes with targeting moieties to selectively be internalized by the cells. The probe would light up differently depending on whether they are within or outside the tumor cells. This difference would assist in identifying areas surrounding the resected tumor. Thus, a more efficient excision would be obtained instead of removing the surrounding healthy tissue simply for precaution. Further, this approach would be useful to identify metastasis that may not be identified using other methods [75]. PEG chains are also commonly used in various biomedical applications to stabilize macromolecules such as proteins and enhance their half-lives. Coarse grained simulations using MARTINI FF in GROMACS were employed to study PAMAM dendrimers of various generations linked to PEG chains of different lengths for studying the conformation/aggregation of the dendrimers in solution [76]. With higher MW PEG chains, a complete coverage of the dendrimer could be achieved and, even though PEG tends to extend towards the water, the spherical characteristics of dendrimers could be maintained. For both lower and higher MW PEG chains, no aggregation between dendrimers was observed, measured by the distance from both the centers of mass [76]. This interesting technique allows studying the impact of surface modifications as well as the prediction of dendrimer behavior to aggregate in solution (obviously making them ineffective).

Using another approach, G5 dendrimers with different end groups, at different pH, were built via Insight II software and simulated via CVFF to correlate the dendrimer conformation with the release/efficacy of an anticancer agent [31]. Seven molecules of 2-methoxyestradiol (2-ME) were randomly incorporated in the simulation box, and the position of 2-ME was measured from the center of mass. The position was observed to be farther for G5 with amine end groups and N-Glycine-OH. Except for G5-carboxyl, all of the other structures exhibited open structures. This was attributed to the release of the drug. The findings were consistent with the lower toxicity of G5-carboxyl noted in cells because a collapsed structure would inhibit the release [31].

4.8 DENDRIMERS INTERACTION WITH LIPID MEMBRANES

Despite the major developments in studying dendrimers, it is still an uphill task to develop novel dendrimers. Irrespective of the applications for which dendrimers may be used in nanomedicine, a fundamental knowledge of how dendrimers interact at the interface with cells is a must. Dendrimers interact with cells, particularly with the lipid bilayer. A number of techniques have been used to study the interaction of dendrimers with biomembranes. These include AFM, DLS, NMR, DSC and Raman Spectroscopy [77–81]. However, it is a formidable task, partly due to the extreme complexity of membranes. The high numbers as well as types of different lipids and proteins modulate the interaction between the dendrimer with the membranes. Polymers with a high density of cationic charge are more likely to cause damaging effects on membranes [82]. Thus, in the case of PAMAM dendrimers toxicity occurred in a concentration, charge density as well as generation-dependent manner [83]. During the interaction, the dendrimer may create pores/holes and disrupt the membrane (increasing with higher generations) or may be well accommodated in the bilayer. Even though this effect may result in a higher cell transfection or might be useful as an antimicrobial agent, it is unsuitable for drug delivery systems [32, 84].

Although there has been great progress in computational sciences, molecular dynamics employing all atoms and explicit solvent systems is limited in terms of time-scale. Coarse-grained models lack a few finer details, e.g., hydrogen bonding but do provide an alternative to complete atomistic simulations. Recent reports suggest that they can successfully be used to build predictive models of dendrimer-membrane interactions [30, 48, 85]. In general, from the various simulation studies, it appears that dendrimers initially interact with lipid membranes through various forces. Then, depending on their composition/size/concentration/membrane properties, different phenomena can occur [86]. Thus, upon interaction with the membrane, no hole formation was observed in case of PAMAM G5 in DMPC membrane [30]. This is in agreement with experimental data where PAMAM G5 could only expand the existing defects but did not create new ones [82]. The distance between the dendrimer and the membrane (on z-axis) can be evaluated to measure dendrimer permeability. Dendrimers G3 and G5 show smaller values which is consistent with the adsorption model for all types of membranes. On the other hand, G7 and G9 values are consistent with the embedded model in the membrane. The permeability also increases with decreasing tail size [30].

Dendrimer concentration is another factor that may increase the toxicity of PAMAMs, at least experimentally. To study this effect on membranes, PAMAM G7 and G9 (on DMPC membrane) were simulated with different simulation box sizes. It was observed that by increasing the area of the membrane, the holes induced by the dendrimer became smaller [30]. However, this effect may not reveal the collective influence of more than one dendrimer at higher concentration since it considers only one dendrimer at a time. To overcome this issue, simulations of both positively charged as well as acetylated dendrimers were tested in clusters in a DMPC bilayer [48]. Interestingly, only 4 positive PAMAM G7 dendrimers were sufficient to induce a firm bending on the membrane with insertion of few branches and pore formation whereas it required 16 positive G5 to induce some bending/insertion in the membrane. Furthermore, these effects were not noted for acetylated G5 dendrimers wherein these dendrimers aggregated [48]. Since the terminal groups may have such impact on the dendrimers interactions, all atom simulations using implicit solvent were carried out. A smaller system of PAMAM G3 dendrimer with different end groups (amine/acetyl/carboxylic) on a DMPC bilayer were used for studying the energy components involved in the interactions. The free energy binding was 47, 36 and 26 kcal/mol (PAMAM-carboxylic/amine/acetyl, respectively). The attractive force was the same for both the charged dendrimers [84]. These results were as per the expectations due to the zwitterion character of DMPC lipid. Thus, charged dendrimers interact more favorably compared to the acetylated neutral ones.

Lipid aggregation may also result in the formation of fluid and gel phases. Experimentally, the AFM measurement of PAMAM G7 revealed that the disrupting mechanism could be abolished in gel phase of DMPC membrane [87]. Coarse-grained simulations of PAMAM G3/G5 in DPPC bilayer at varying temperature (277 K and 310 K) to simulate a better-ordered phase in comparison to a disordered phase. It was noted that during the simulation time, no insertion of PAMAM G5 occurred at 277 K, in contrast to the insertion observed at 310 K [18]. Likewise, all atom simulations of PAMAM dendrimers terminated differently (amine/carboxyl/acyl groups) with implicit solvent treatment. The gel phase was simulated by immobilizing the lipid tails from an equilibrated DMPC bilayer in fluid phase. In case of the fluid membrane, a depression was created to accommodate the dendrimer whilst the dendrimer flattened out to extend the number of interactions. On the other hand, in the gel phase, the dendrimers remained at the surface and did not induce any kind of deformity. However, cell membranes may be composed of different lipids and the various combinations may lead to different properties. To address this point, a study using coarse-grained description of 3 types of lipids (DPPC/DPPE/DPPS) tested different ratios (mimicking erythrocyte membrane) with G4/G5 PAMAMs [19]. The asymmetry of the membranes was correlated with the ability of the dendrimers to get insert on it. In case of symmetric membranes, PAMAM G4 remained on the surface of the outer leaflet. These differences were explained on the basis of electrostatic attraction between the inner leaflets of the membranes for the dendrimer. Further, increasing the percentage of DPPS (from 10% to 50%) showed a fall in the distance between center of masses of the dendrimer and the membrane. This implies that the insertion of the dendrimer was greater. Complementarily, order parameter of phospholipid tails (which measures the movements of lipid bilayers) revealed that the incorporation of the dendrimer result in decreased lipid order. PAMAM G5 created more alterations in the lipid order that caused a transient formation of a pore [19]. The chemical nature of membrane lipids can also have a role in the interaction with dendrimers. By using coarse-grained simulations of PAMAM G3/G4 dendrimer in bilayers made up of DPPC mixed with dipalmitoyl phosphatidylglycerol (DPPG), it was observed that the dendrimers enhanced the diffusion of oppositely charged DPPG from the inner to the outer leaflet [88, 89]. This specific formation of lipid raft microdomains was also noted for PAMAM G7/G8/G9 interacting with mixture (bilayer) of 1,2-dimyristoyl-glycero-3-phosphoglycerol (DMPG) with DMPC by using coarse-grained simulations employing the MARTINI force field. The simulation revealed that the number of DMPG lipids close to the dendrimer went up with an increase in dendrimer generation and a significant turn in the membrane was observed [90]. Overall, such simulations are consistent with the experimental data on the toxicity of dendrimers. The specific interest in performing molecular dynamics is not merely to explain how these effects take place, but also to prevent them by designing such structures which do not favor these interactions. In particular, it would be of special interest to develop dendrimers that would efficiently include a mechanism of lysosomal escape without the disruptive mechanism. This should be useful for a wide range of therapeutic molecules including gene delivery.

4.9 MODELING DENDRIMERS FOR DRUG DELIVERY APPLICATIONS

Combining the power of explaining molecular mechanisms and to construct predictive models that could be applied to various kinds of dendrimers, yields a major advantage to carry out optimizations even before commencing the synthesis of novel dendrimers. Speaking in terms of drug delivery applications, there are certain factors where modeling as well as molecular dynamics provide useful insights on the way to optimize these carriers (Table 4.8).

The detailed analysis of these factors is also crucial to understand how these dendrimers will perform *in vivo*. In the field of drug delivery designing, dendrimers can be designed and catered to a vast range of drugs in

TABLE 4.8 Factors Influencing Optimization of Dendrimers as Drug Carriers

- Stability of drug-dendrimer complexes.
- Strength of interactions that might compromise drug release.
- Availability of the targeting groups for interaction.
- Exposure of labile molecules to the solvent.
- Elucidate which forces govern the dendrimer-drug interaction.
- Whether the above forces can be changed.
- Ability of the dendrimer to incorporate a significant number of drug molecules.

order to solve issues such as solubility, drug release and drug targeting. Essentially, drugs may be incorporated into the dendrimer either covalently (e.g., a prodrug) or non-covalently (on the surface or in the internal cavities). As observed in case of the structural behavior of PAMAM dendrimers, the hydrophobicity of the core may be modulated via pH and ionic variations. These alterations can be used to determine the conditions for drug encapsulation that would be different from the medium from which they would be released.

A classical example is the delivery of NSAIDs (non-steroidal anti-inflammatory drugs). The special interest in these drugs lies in their promising potential for encapsulation as well as controlled release. This is likely to be technologically useful as a prolonged analgesic effect would be achieved without increasing the adverse effects such as gastrointestinal ulceration. Ibuprofen is a common model drug for testing new drug delivery systems owing to its small size and being a class II drug molecule (high permeability with low solubility) so that its bioavailability is limited by the solvation rate. Experiments have shown that PAMAM dendrimers solubilize ibuprofen and this effect was dependent on pH as well as the generation of PAMAM Dendrimers [91, 92]. To find out how the interaction occurs, a PAMAM G3 was simulated at another pH with ibuprofen [12]. Using all-atom MD simulation with AMBER FF, the drug was placed in the proximity of the PAMAM dendrimer. The analysis of atom distributions made it clear that ibuprofen could penetrate more towards the dendrimer core than water molecules. Unlike at basic or neutral pH, at an acidic pH, ibuprofen located homogenously throughout the surface of dendrimer. However, cluster formation was observed. The measurement of the mean distance between the center of mass of ibuprofen and PAMAM (as a function of time) also suggested a constant value at neutral/basic pH. This means that the complexes formed are stable. At neutral pH, hydrogen bonds are formed between the dendrimer and the ibuprofen so that this complex is largely formed at the surface. At acidic pH, on the contrary, ibuprofen diffuses away as confirmed by the increased values of average distance [12]. This finding was consistent with the experimental results wherein the solubility of PAMAM dendrimers falls under acidic conditions [91]. In another study, calculation of dendrimer-ibuprofen complexes with encapsulation of different amounts of drugs was assessed using MM2 calculations. From these data it was observed that while incorporating more than 16 ibuprofen molecules inside the dendrimer, the energetics were quite unfavorable, and this is a similar value to the one determined experimentally (with 14 ibuprofen molecules) for PAMAM G3 [59, 92]. These kinds of studies can also be applied to other drugs of the same class to investigate which drugs show better binding affinity to the dendrimers. Thus, a predictive model can be constructed for the drug better suited for the carrier. In this case, the designing is based on the selection of the drug and not on the optimization of the carrier. In order to achieve a more accurate data on the binding energies of four different NSAIDs, semi-empirical methods (using PM6-DH+) were employed [93]. However, this approach is restricted by the size of the dendrimer to be considered in the calculations. To overcome this problem, some of the branches of the dendrimer were picked up separately from structures arising from MD simulations. The conformational pairs formed between the branches and the drugs were subsequently generated by the Monte Carlo method and the energetics were calculated using semi-empirical methods. The energy values that were obtained could be directly correlated with the degree of affinity determined experimentally (they were- naproxen>ketoprofen>ibuprofen> diflunisal) [93]. In a similar kind of study, but applicable for anti-cancer therapy, MD simulations were performed to assess the molecular interactions between a series of 24 anticancer drugs and a G4 PAMAM dendrimer. The results indicate that most of the drugs demonstrate high thermodynamic stability. The drugs effectively interacted with the dendrimer in an exothermic manner, with bleomycin, orlistat and porfimer most strongly interacting with the PAMAM dendrimer [94]. From these encouraging results, one can predict from a pool of candidate drugs which ones would best fit into the parameters of a given dendrimer. This kind of approach was tested on PAMAM G5 dendrimers with other drugs (L-alanine, salicylic acid, primidone and phenylbytazone) [45]. The drugs were docked to the dendrimer through AutoDock Vina and the best scoring conformations were chosen for MD simulations by using AMBER with an explicit water solvent. Samplings were carried out between the centers of masses of the dendrimer/drugs. While plotting the PMF (potential mean force) among all the drugs, L-alanine showed lower free energy (which translates into a better ability to be released) followed by salicylic acid then primidone and finally phenylbutazone. However, considering the experimental data, even though L-alanine/salicylic acid had lower free energies, it was difficult to encapsulate them in the dendrimer due to lack of nonpolar groups. This, in turn, is due to lower van der Waals contributions and hydrogen bonding that did not significantly contribute to the free energy barrier. The PMF was also lower when the drugs are linked to the non-protonated dendrimer. The authors therefore suggest that drugs should be encapsulated at a higher pH; once at physiological pH, they would be more tightly bound and enable the release to be controlled [43]. Likewise, PAMAM G4 equilibration was studied by using MM+ FF and docked with resveratrol, genistein as well as curcumin [13]. The MD simulations showed that the free energy binding followed the order genistein>curcumin>resveratrol with different binding constants determined experimentally (being curcumin> genistein>resveratrol). This difference between the estimation of energy of binding and the binding constants could be attributed to the constraints of the drug to access the core of the dendrimer. A way of circumventing this kind of behavior would be by employing adaptive biasing force methods. Such an approach has been recently used to evaluate the association of nicotinic acid and PAMAM G3 at varying pH (pH 3 and 6) [41]. By using a biasing force method, the drug molecules were constrained in z-axis to be sure that the drug travelled along the selected sample direction though allowed to travel freely through x and y axes. The energy profiles showed nicotinic acid to interact better with G3 at higher pH values, with an energy gap of −1 to −2.5 kcal in contrast to 3-pyridinium carboxylate. This interaction was favorable more at the surface rather than in the interior [41]. Since nicotinic acid is poorly soluble and should be delivered to the interior of cells, determining the conditions at which it can be better encapsulated in the dendrimer as well as its location, may serve as a guide for optimization of the carrier in future.

A different approach was used to evaluate the stability of the complex formation between poly(L-lysine) G6 dendrimers and an anticancer drug, doxorubicin, in order to evaluate its potential as a drug delivery system. The drug-dendrimer complex was favorable at 300 K but dissociated on heating till 1000 K. However, once the system got cooled, it reassembled and again showed the favorable interaction between the two molecules suggesting its favorable potential for controlled release [44].

One of the most interesting and promising areas for the use of dendrimers as drug delivery tools is gene delivery. Dendrimers are known potential carriers for nucleic acids owing to their high positive charge that allows them to form dendriplexes (polyelectrolyte complexes). Experimental as well as all-atom MD simulations have demonstrated that nucleic acids are able to wrap around the dendrimers in a process depending both on size and charge ratio. MD simulations can probe with reasonable detail whether a dendrimer could be a suitable carrier. MD simulations were performed in AMBER7 (with AMBER95 force field for ssDNA and the DREIDING force field for the dendrimer) with varying levels of protonation. The dendrimers were docked to the major groove of ssDNA. Further simulations were performed in explicit water solvent and counter ions were added in order to neutralize the system. The simulations revealed that for G2 and G3 the charge ratio was inadequate to fully wrap ssDNA onto the dendrimers, as suggested by the Rg of the complexes. On the other hand, G4 dendrimer being large enough, it neutralized the charges of ssDNA and enhanced the latter's collapse latter onto the surface. This led to a compact complex formation with major penetration inside as assessed by the radial density distribution. At neutral pH, ssDNA better penetrated inside the dendrimer. However, this may not necessarily be advantageous since it might prevent the release from dendrimer, thereby not rendering it useful as a gene delivery vehicle [49]. In a similar way, protonated PAMAM G3 and G4 dendrimers linked with ssDNA were simulated using AMBER03 force field for ssDNA, GAFF for the linker and DREIDING force field for the dendrimer. Once again, ssDNA lost helicity and failed to wrap around the dendrimer. There was higher wrapping and DNA penetration with G4 dendrimer [95]. Even when B-form of dsDNA was simulated with PAMAM G3 to G5 dendrimers, a strong deformation of DNA was noted [96]. Performing MD simulations with AMBER03 force field in explicit water solvent and added ions, G5 expanded in an attempt to cover the entire DNA as the DNA wrapped around the dendrimer as measured by the Rg [96]. At the initial stage of the complex formation, the dendrimer expanded while increasing the contact between the DNA and dendrimer. Water molecules then underwent repulsion from the DNA backbone structure and DNA wrapped around the dendrimer, forming a more stable complex. However, this phenomenon is probably limited to the number of generation because it could be shown by the MD of G7 PAMAM dendrimers with siRNA wherein the dendrimer behaved as a rigid sphere with no variation in Rg post-binding [47]. Higher charge ratio translates into higher binding interaction [96]. Although G3 is not enough to neutralize DNA and a weak interaction does occur, this could be a better system since the release should be easier than what it would be for G4 or G5 [96]. From this aspect, different generations of flexible triazine dendrimers and PEI polymer were simulated using DsiRNA in AMBER (parm99 force field) in order to predict their efficacy [97]. Thermodynamically, dendrimers were noted to be more stable than PEI. G2 was the most stable complex followed by G4 and then G3. Further, charge neutralization in 1:1 complex predicted the stability of the complexes in solution since it was hypothesized that PEI only partly interacted with the DsiRNA. It was suggested that the non-complexed part bearing both positive and negative charges enhanced inter-particle interactions and led to aggregation. Finally, further studies showed that G4 was more stable in contrast to G2 (reduced excretion) but underwent significantly more uptake by the reticuloendothelial system [97].

The DNA sequence also plays a role in the dendrimer-nucleic acid complexation and herein lies the importance of making use of computational methods to predict their interaction. By using MD simulations with varying strand composition it has been found that the binding constant follows the order as polyG>polyC>polyA>polyT, as revealed by the free energy calculations [46]. The flexibility/rigidity of dendrimers is another critical point in the formation of the polyelectrolyte complexes which is due to a balance between the enthalpy and entropy of binding [45].

Dendrimer-based systems have the potential to be used as contrast agents because they can modulate the pharmacokinetics and organ selection. Gadolinium-based triazine dendrimers bearing DOTA chelate groups were studied as contrast agents by using MD simulations with AMBER force field. In this model, G3/G5 dendrimers possessed 24 and 96 chelates respectively [37]. Analysis of the radial distribution functions suggested that the chelates were exposed to solvent and were available for Gd ions chelation. In fact, the tall peaks in the RDF suggested a reduced back-folding throughout the entire simulation time.

The multivalence of dendrimers is one of their topmost regarded merits in which they can be designed to perform multiple functions within the same structural boundary. This is especially useful in the application with imaging agents. In this regard, a G5 PAMAM linked with gold nanoparticles and with randomly distributed folic acid as targeting ligand, and fluorescein isothiocyanate as imaging agent, were simulated using CVFF [38]. The MD simulation was quite useful in revealing that the fluorescent probe was at a significant distance from gold. Therefore, a low quenching was expected but the folic acid extended outwards and made it available to interact. This system is definitely interesting since it is an 'all-in-one package.' Further, this dendrimer-based system can selectively target cancer, highlighting its location, so that finally laser hyperthermia may be induced on the nanoparticle.

4.10 MODELING DENDRIMERS AS THERAPEUTIC AGENTS

Dendrimers have an inherent similarity to bio-macromolecules such as proteins and may interact with other biomolecules to illicit biological responses [16, 62]. The large number of terminal groups promotes multivalent and non-specific interactions that are beneficial. Furthermore, specific modifications may be introduced at the surface, which promote binding. By using simulations on G0 PAMAM modified with guanidinium, it was observed that these modifications enhanced simultaneous and cooperative interactions with the surface of α-chymotrypsinogen A via hydrogen and cationic interactions with the amino acids [98]. This cooperative interaction promoted higher binding when compared to a single unit.

Dendrimers can be designed and modeled for specific interactions with the desired target. G3.5 PAMAM dendrimers have been modeled to predict the surface availability of glucosamine moieties and their influence on biological activity. Experimentally, PAMAM G3.5 (having carboxylic groups) with an average number of 8 surface glucosamine molecules inhibited the formation of TLR5-MD-2-LPS complex [99]. Frontier molecular orbital theory (FMOT) was applied on the smaller Dendrimer dimensions for predicting the reactivity (the gap between HOMO and LUMO) of the Dendrimer for later addition of glucosamine monomers [36]. The studies found that based solely on the electronic properties of the dendrimer, not more than 12 molecules can be coupled to the structure. However, upon adding more glucosamine, the steric effects of hindering the end groups began to occur as evidenced by the MD simulations. Taken together, these effects accounted for an average of 4 to 8 as the most favorable, energetically [36]. The dendrimers were then docked to MD-2 protein followed by MD simulation. It was found that the dendrimer can indeed behave as an MD-2 receptor antagonist [16] (Figure 4.2). Analysis of the trajectory revealed that the PAMAM-glucosamine groups can cooperatively bind to hydrophilic residues located at the entry to the hydrophobic pocket [16]. This action was enough to prevent the entrance of LPS and the effect was important since it was shown to prevent TLR4-MD-MD-2-LPS from triggering the cytokine cascade. The predictive knowledge of

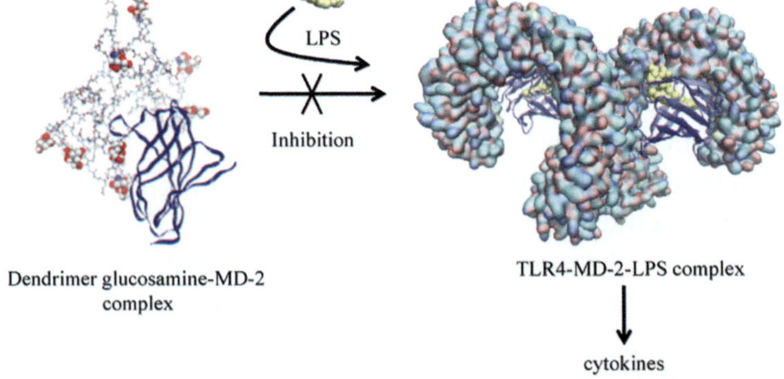

FIGURE 4.2 Docking between dendrimer glucosamine and MD-2 protein, and inhibition of TLR4-MD-2-LPS complex formation. (From Martinho, N.; Florindo, H.; Silva, L.; Brocchini, S.; Zloh, M.; Barata, T. Molecular Modeling to Study Dendrimers for Biomedical Applications. Molecules 2014, 19, 20424-20467. https://creativecommons.org/licenses/by/4.0/)

the dendrimer-glucosamine derivative/receptor was exploited as the basis for further designing the dendrimer architecture. Active as well as inactive PETIM [100] dendrimers and Triazine-PAMAM hybrids [9] were modeled using the same basic principles. Modified PETIM dendrimers showed the same flexibility and net surface charge was resembled that of the PAMAM derivative. This dendrimer was especially interesting since it was a lower molecular weight polymer as compared to the original PAMAM dendrimer (besides easier synthesis and purification). The results from inactive forms of the dendrimer suggested that surface hydrophobicity had no contribution towards the interaction with MD-2 receptor. Similarly, in case of the Triazine-PAMAM hybrids [9] results showed that the best architecture that promoted antagonist effect was that of partly glycosylated hybrid G2 Triazine/G1.5 PAMAM dendrimers. This dendrimer resembled active G3.5 PAMAM-glucosamine in certain key points such as the polar surface area and the globular shape (which is contrary to the pure Triazine). These features promote electrostatic interactions with the target and might justify its activity.

In the biomedical field, another interesting application of dendrimers is in the domain of vaccination where a core-based dendrimer may be used to attach multiple moieties depending on the type of vaccination required. Using a database of *P. falciparum* epitopes, many epitopes were selected and linked to the dendrimer [101]. These systems were simulated by using CHARMM force field and explicit water solvent, and energetics of the different structures was determined. This is a potentially interesting application of MD simulations since one can estimate the availability of epitopes to the solvent. It can further be used to observe the effect of adding other groups, e.g., recognition groups or fluorophores in order to track the intracellular pathway of these dendrimers. Although dendrimers may be designed to specifically interact with proteins, while circulating (i.e., *in vivo*) they may contact other proteins and cause significant alteration of the pharmacokinetics. A common test has been to study the binding affinity with albumin which is the most abundant plasma protein in humans. In particular, dendrimers with the high number of end groups may promote nonspecific binding and thus bind to plasma proteins. Molecular modeling offers a potential means to evaluate as well as modulate these interactions. Using the DREIDING force field (which predicts the behavior of PAMAM at varying pH), PAMAM dendrimers have been simulated with human serum albumin (HSA) to assess the contacts between the two. It was reported that the size/surface of end groups were critical for this interaction. Further, the interaction between dendrimer-HSA was attributed to electrostatic interactions between charged groups besides hydrogen bonding and hydrophobic interactions, and resulted in a back-folding of PAMAM dendrimers (inner groups participating in the interaction) [102]. From the SASA calculations for estimating the pertinent contact areas using various probe radii size, and the NMR data, it was proposed that PAMAM forms weak complexes with the HSA [102].

4.11 NOW AND NEXT?

At present, there are more than 100,000 entries in the protein databank depicting the experimentally determined structures of numerous biologically relevant molecules. In contrast, very few entries exist showing the structures of dendrimers. This clearly indicates that the intrinsic properties of dendrimers prevent the experimental elucidation of their architecture as well as their interactions with biological components. Computational/molecular modeling and dynamics are well known valuable tools for conventional drug designing and optimization, and may be successfully applied to study dendrimers. Molecular modeling when intelligently applied to dendrimers would offer a means to study their conformation and multiple features of dendrimers at a molecular level that are otherwise difficult to study experimentally. Such approaches would also elucidate the key interactions between functionalized dendrimers and therapeutic molecules or biological systems, e.g., proteins and lipid membranes. There are at present two key challenges as far as the modeling of dendrimers is concerned. The first is obtaining the initial 3-dimensional coordinates of the dendrimer that are to be used in the subsequent computational studies. This can be manually performed by sketching the dendrimer with common tools of chemical drawing. This is relatively easy for small generation dendrimers. However, with larger generation dendrimers, the approach is error-prone [29, 56]. There are some major tools that can generate the

TABLE 4.9 Major Tools to Generate the Starting 3-Dimensional Coordinates of a Dendrimer

Tool	Specifications
Dendrimer Builder Toolkit [24]	A graphical user interface, written in PERL, interfacing with AMBERTOOLS
Silico [23]	Written in Perl, has a module called Starmaker, generates mol2 dendrimer files by assembling them layer by layer.
Python GUI [25]	Using XPLOR, defining the sequence of monomers and the manner in which they are connected to each other via patch references. The structures generated using such an approach may be converted to file formats having generalized force fields from XPLOR.

starting 3-dimensional coordinates of a dendrimer under consideration (Table 4.9).

The other major issue while computationally simulating dendrimer architectures is the lack of dedicated force fields. This is perhaps due to the wide range of chemical linkages used in assembling dendrimers the fact that there are only a few reported crystal structures of dendrimers. Because of this, studies very often rely on the force fields that have been developed for other molecules like proteins. Developing more generalized atom types in the force fields would permit the automated generation of dendrimers. This would help in determining structure-property-functional correlations needed to gain insight into their behavior. Further, these structures can be used in docking experiments. The docking solutions can also be used in the MD simulations to explore in details the intermolecular interactions. Docking studies also permit the selection of the optimum dendrimer architecture for synthesizing and further studies on optimization. Hopefully, sufficient MD data about dendrimer types would become available in future. The continued computation studies would yield new analytical methods for their dynamic structural properties which is essential for their design/optimization. Finally, the refinement and continued evolution of computational power would allow the simulation times to explore more complex interactions.

KEYWORDS

- **computational modeling**
- **dendrimer topology**
- **drug delivery**
- **flexible dendrimers**
- **lipid membranes**
- **therapeutic agents**

REFERENCES

1. Caminade, A.-M., Laurent, R., & Majoral, J.-P. (2005). Characterization of dendrimers. *Adv. Drug Deliv. Rev. 57,* 2130–214.
2. Cagin, T., Wang, G., Martin, R., Zamanakos, G., Vaidehi, N., Mainz, T., & Iii, W. A. G. (2001). Multiscale modeling and simulation methods with applications to dendritic polymers. *Comput. Theor. Polym. Sci. 11*, 345–356.
3. Lee, H., & Larson, R. G. (2009). Multiscale modeling of dendrimers and their interactions with bilayers and polyelectrolytes. *Molecules 14*, 423–438.
4. Ranganathan, D., & Kurur, S. (1997). Synthesis of totally chiral, multiple armed, poly Glu and poly Asp scaffoldings on bifunctional adamantane core. *Tetrahedron Lett. 38*, 1265–1268.
5. Zloh, M., Ramaswamy, C., Sakthivel, T., Wilderspin, A., & Florence, A. T. (2005). Investigation of the association and flexibility of cationic lipidic peptide dendrons by NMR spectroscopy. *Magn. Reson. Chem. 43*, 47–52.
6. Tian, W., & Ma, Y. (2009). Molecular dynamics simulations of a charged dendrimer in multivalent salt solution. *J. Phys. Chem. B 113*, 13161–13170.
7. Ouyang, D., Zhang, H., Parekh, H. S., & Smith, S. C. (2011). The effect of pH on PAMAM dendrimer-siRNA complexation: Endosomal considerations as determined by molecular dynamics simulation. *Biophys. Chem. 158*, 126–133.
8. Jain, V., Maingi, V., Maiti, P. K., & Bharatam, P. V. (2013). Molecular dynamics simulations of PPI endrimer–drug complexes. *Soft Matter 9*, 6482–6496.
9. Barata, T. S., Teo, I., Lalwani, S., Simanek, E. E., Zloh, M., & Shaunak, S. (2012). Computational design principles for bioactive dendrimer based constricts as antagonists of the TLR4-MD-2-LPS complex. *Biomaterials 32*, 8702–8711.
10. Barnard, A., Posocco, P., Pricl, S., Calderon, M., Haag, R., Hwang, M. E., Shum, V. W. T., Pack, D. W., Smith, D. K. (2011). Degradable self-assembling dendrons for gene delivery: Experimental and theoretical insights into the barriers to cellular uptake. *J. Am. Chem. Soc. 133*, 20288–20300.

11. Javor, S., & Reymond, J.-L. (2009). Molecular dynamics and docking studies of single site esterase peptide dendrimers. *J. Org. Chem. 74,* 3665–3674.
12. Tanis, I., & Karatasos, K. (2009). Association of a weakly acidic anti-inflammatory drug (ibuprofen) with a poly(amidoamine) dendrimer as studied by molecular dynamics simulations. *J. Phys. Chem. B 113,* 10984–10993.
13. Abderrezak, A., Bourassa, P., Mandeville, J.-S., Sedaghat-Herati, R., & Tajmir-Riahi, H.-A. (2012). Dendrimers bind antioxidant polyphenols and cisplatin drug. *PLoS One 7,* e33102.
14. Brocchini, S., Godwin, A., Balan, S., Choi, J., Zloh, M., & Shaunak, S. (2008). Disulfide bridge based PEGylation of proteins. *Adv. Drug Deliv. Rev. 60*, 3–12.
15. Uhlich, N. A., Darbre, T., & Reymond, J.-L. (2011). Peptide dendrimer enzyme models for ester hydrolysis and aldolization prepared by convergent thioether ligation. *Org. Biomol. Chem. 9*, 7071–7084.
16. Barata, T. S., Teo, I., Brocchini, S., Zloh, M., & Shaunak, S. (2011). Partially Glycosylated Dendrimers Block MD-2 and Prevent TLR4-MD-2-LPS Complex Mediated Cytokine Responses. *PLoS Comput. Biol. 7*, e1002095.
17. Monticelli, L., Kandasamy, S. K., Periole, X., Larson, R. G., Tieleman, D. P., & Marrink, S.-J. (2008). The MARTINI Coarse-Grained Force Field: Extension to Proteins. *J. Chem. Theory Comput. 4*, 819–834.
18. Lee, H., & Larson, R. G. (2006). Molecular dynamics simulations of PAMAM dendrimer-induced pore formation in DPPC bilayers with a coarse-grained model. J. *Phys. Chem. B 110*, 18204–18211.
19. He, X., Qu, Z., Xu, F., Lin, M., Wang, J., Shi, X., & Lu, T. (2014). Molecular analysis of interactions between dendrimers and asymmetric membranes at different transport stages. *Soft Matter 10*, 139–148
20. Lozac'h, N., Goodson, A. L., & Powell, W. H. (1979). Nobel Nomenclature- General Principles. *Angew. Chem. Int. Ed. Engl. 18*, 887–899.
21. Hess, B., Kutzner, C., van der Spoel, D., & Lindahl, E. (2007). GROMACS 4: Algorithms for Highly Efficient, Load-Balanced, and Scalable Molecular Simulation. *J. Chem. Theory Comput. 4,* 436–447.
22. Schwieters, C., Kuszewski, J., & Mariusclore, G. (2006). Using Xplor-NIH for NMR molecular structure determination. *Proc. Nucl. Magn. Reson. Spectrosc. 48,* 47–62.
23. Charlmers, D., & Roberts, B. Silico—A Perl Molecular Modeling Toolkit. Available online: http://silico.sourceforge.net/Silico/Home.html (accessed on 1 August 2014).
24. Maingi, V., Jain, V., Bharatam, P. V., & Maiti, P. K. (2012). Dendrimer building toolkit: Model building and characterization of various dendrimer architectures. *J. Comput. Chem. 33*, 1997–2011.
25. Barata, T. S., Brocchini, S., Teo, I., Shaunak, S., Zloh, M. (2011). From sequence to 3D structure of hyperbranched molecules: Application to surface modified PAMAM dendrimers. *J. Mol. Model. 17*, 2741–2749.
26. Pavan, G. M., Albertazzi, L., & Danani, A. (2010). Ability to adapt: Different generations of PAMAM dendrimers show different behaviors in binding siRNA. *J. Phys. Chem. B 114*, 2667–2675.
27. Lim, J., Lo, S.-T., Hill, S., Pavan, G. M., Sun, X., & Simanek, E. E. (2012). Antitumor activity and molecular dynamics simulations of paclitaxel-laden triazine dendrimers. *Mol. Pharm. 9*, 404–412.

28. Mills, M., Orr, B. G., Banaszak Holl, M. M., & Andricioaei, I. (2013). Attractive hydration forces in DNA-dendrimer interactions on the nanometer scale. *J. Phys. Chem. B 117*, 973–981.
29. Filipe, C. S., Machuqueiro, M., Darbre, T., & Baptista, M. (2013). Unraveling the Conformational Determinants of Peptide Dendrimers Using Molecular Dynamics Simulations. *Macromolecules 46*, 9427–9436.
30. Wang, Y.-L., Lu, Z.-Y., & Laaksonen, A. (2012). Specific binding structures of dendrimers on lipid bilayer membranes. *Phys. Chem. Chem. Phys. 14,* 8348–8359.
31. Shi, X., Lee, I., Chen, X., Shen, M., Xiao, S., Zhu, M., Baker, J. R., Jr., & Wang, S. H. (2010). Influence of dendrimer surface charge on the bioactivity of 2-methoxyestradiol complexed with dendrimers. *Soft Matter 6,* 20–27.
32. Stach, M., Maillard, N., Kadam, R. U., Kalbermatter, D., Meury, M., Page, M. G. P., Fotiadis, D., Darbre, T., & Reymond, J.-L. (2012). Membrane disrupting antimicrobial peptide dendrimers with multiple amino termini. *Med. Chem. Comm. 3*, 86–89.
33. Miklis, P., Tahir, C., & Iii, W. A. G. (1997). Dynamics of Bengal Rose Encapsulated in the Meijer Dendrimer Box. *J. Am. Chem. Soc. 119*, 7458–7462.
34. Lee, I., Athey, B. D., Wetzel, A. W., Meixner, W., & Baker, J. R. (2002). Structural Molecular Dynamics Studies on Polyamidoamine Dendrimers for a Therapeutic Application: Effects of pH and Generation. *Macromolecules 35*, 4510–4520.
35. Lee, H., Baker, J. R., & Larson, R. G. (2006). Molecular dynamics studies of the size, shape, and internal structure of 0% and 90% acetylated fifth-generation polyamidoamine dendrimers in water and methanol. *J. Phys. Chem. B 110*, 4014–4019.
36. Barata, T. S., Shaunak, S., Teo, I., Zloh, M., & Brocchini, S. (2011). Structural studies of biologically active glycosylated polyamidoamine (PAMAM) dendrimers. *J. Mol. Model. 17*, 2051–2060.
37. Lim, J., Turkbey, B., Bernardo, M., Bryant, L. H. J., Garzoni, M., Pavan, G. M., Nakajima, T., Choyke, P. L., Simanek, E. E., & Kobayashi, H. (2012). Gadolinium MRI contrast agents based on triazinedendrimers: Relativity and in vivo pharmacokinetics. *Bioconjug. Chem. 23*, 2291–2299.
38. Shi, X., Wang, S., Meshinchi, S., van Antwerp, M. E., Bi, X., Lee, I., & Baker, J. R. (2007). Dendrimer-Entrapped Gold Nanoparticles as a Platform for Cancer-Cell Targeting and Imaging. *Small 3*, 1245–1252.
39. Tyssen, D., Henderson, S. A., Johnson, A., Sterjovski, J., Moore, K., La, J., Zanin, M., Sonza, S., Karellas, P., Giannis, M. P., et al. (2010). Structure Activity Relationship of Dendrimer Microbicides with Dual Action Antiviral Activity. *PLoS One 5*, e12309.
40. Roberts, B. P., Krippner, G. Y., Scanlon, M. J., & Chalmers, D. K. (2009). Molecular Dynamics of Variegated Polyamide Dendrimers. *Macromolecules 42*, 2784–2794.
41. Caballero, J., Poblete, H., Navarro, C., & Alzate-Morales, J. H. (2013). Association of nicotinic acid with a poly (amidoamine) dendrimer studied by molecular dynamics simulations. *J. Mol. Graph. Model. 39*, 71–78.
42. Carrasco-Sánchez, V., Vergara-Jaque, A., Zuñiga, M., Comer, J., John, A., Nachtigall, F. M., Valdes, O., Duran-Lara, E. F., Sandoval, C., & Santos, L. S. (2014). In situ and in silico evaluation of amine- and folate-terminated dendrimers as nanocarriers of anesthetics. *Eur. J. Med. Chem. 73*, 250–257.

43. Maingi, V., Kumar, M. V. S., & Maiti, P. K. (2012). PAMAM dendrimer-drug interactions: Effect of pH on the binding and release pattern. *J. Phys. Chem. B 116*, 4370–4376.
44. Al-Jamal, K. T., Al-Jamal, W. T., Wang, J. T.-W., Rubio, N., Buddle, J., Gathercole, D., Zloh, M., & Kostarelos, K. (2013). Cationic poly-L-lysine dendrimer complexes doxorubicin and delays tumorgrowth in vitro and in vivo. *ACS Nano* 7, 1905–1917.
45. Pavan, G. M., Mintzer, M. A., Simanek, E. E., Merkel, O. M., Kissel, T., & Danani, A. (2010). Computational insights into the interactions between DNA and siRNA with "rigid" and "flexible" triazine dendrimers. *Biomacromolecules 11*, 721–730.
46. Maiti, P. K., & Bagchi, B. (2006). Structure and dynamics of DNA-dendrimer complexation: Role of counterions, water, and base pair sequence. *Nano Lett. 6,* 2478–2485.
47. Jensen, L. B., Mortensen, K., Pavan, G. M., Kasimova, M., Jensen, D. K., Gadzhyeva, V., Nielsen, H. M., & Foged, C. (2010). Molecular Characterization of the Interaction between siRNA and PAMAM G7 Dendrimers by SAXS, ITC, and Molecular Dynamics Simulations. *Biomacromolecules 11*, 3571–3577.
48. Lee, H., & Larson, R. G. (2008). Coarse-grained molecular dynamics studies of the concentration and size dependence of fifth- and seventh-generation PAMAM dendrimers on pore formation in DMPC bilayer. *J. Phys. Chem. B 112,* 7778–7784.
49. Kelly, C. V., Leroueil, P. R., Orr, B. G., Banaszak Holl, M. M., & Andricioaei, I. (2008). Poly(amidoamine) dendrimers on lipid bilayers II: Effects of bilayer phase and dendrimer termination. *J. Phys. Chem. B 112*, 9346–9353.
50. Zhong, T., Ai, P., & Zhou, J. (2011). Structures and properties of PAMAM dendrimer: A multi-scale simulation study. *Fluid Phase Equilib. 302*, 43–47.
51. Prosa, T. J., Bauer, B. J., Amis, E. J., Tomalia, D. A., & Scherrenberg, R. (1997). A SAXS study of the internal structure of dendritic polymer systems. *J. Polym. Sci. Part B Polym. Phys. 35*, 2913–2924.
52. Liu, Y., Bryantsev, V. S., Diallo, M. S., & Goddard, W. A. (2009). PAMAM dendrimers undergo Ph responsive conformational changes without swelling. *J. Am. Chem. Soc. 131*, 2798–2799.
53. Antosiewicz, J. M., & Shugar, D. (2011). Poisson-Boltzmann continuum-solvation models: Applications to pH-dependent properties of biomolecules. *Mol. Biosyst. 7*, 2923–2949.
54. Ballauff, M., & Likos, C. N. (2004). Dendrimers in solution: Insight from theory and simulation. *Angew. Chem. Int. Ed. Engl. 43,* 2998–3020.
55. Lyulin, S. V., Darinskii, A. A., & Lyulin, A. V. (2005). Computer Simulation of Complexes of Dendrimers with Linear Polyelectrolytes. *Macromolecules 38,* 3990–3998.
56. Filipe, L. C. S., Machuqueiro, M., & Baptista, A. M. (2011). Unfolding the conformational behavior of peptide dendrimers: Insights from molecular dynamics simulations. *J. Am. Chem. Soc. 133,* 5042–5052.
57. Barra, P. A., Barraza, L., Jiménez, V. A., Gavín, J. A., & Alderete, J. B. (2014). Complexation of Mefenamic Acid by Low-Generation PAMAM Dendrimers: Insight from NMR Spectroscopy Studies and Molecular Dynamics Simulations. *Macromol. Chem. Phys. 215*, 372–383.

58. Quintana, A., Raczka, E., Piehler, L., Lee, I., Myc, A., Majoros, I., Patri, A. K., Thomas, T., Mulé, J., & Baker, J. R. (2002). Design and function of a dendrimer-based therapeutic nanodevice targeted to tumor cells through the folate receptor. *Pharm. Res. 19,* 1310–1316.
59. Evangelista-Lara, A., & Guadarrama, P. (2005). Theoretical evaluation of the nanocarrier properties of two families of functionalized dendrimers. *Int. J. Quantum Chem. 103*, 460–470.
60. Cao, J., Zhang, H., Wang, Y., Yang, J., & Jiang, F. (2013). Investigation on the interaction behavior between curcumin and PAMAM dendrimer by spectral and docking studies. *Spectrochim. Acta A Mol. Biomol. Spectrosc. 108,* 251–255.
61. Vergara-Jaque, A., Comer, J., Monsalve, L., González-Nilo, F. D., & Sandoval, C. (2013). Computationally efficient methodology for atomic-level characterization of dendrimer-drug complexes: A comparison of amine- and acetyl-terminated PAMAM. *J. Phys. Chem. B 117*, 6801–6813.
62. Maiti, P. K., Ça□ın, T., Lin, S.-T., & Goddard, W. (2005). A. Effect of Solvent and pH on the Structure of PAMAM Dendrimers. *Macromolecules 38*, 979–991.
63. Jain, K., Kesharwani, P., Gupta, U., & Jain, N. K. (2010). Dendrimer toxicity: Let's meet the challenge. *Int. J. Pharm. 394*, 122–142.
64. Chen, W., Porcar, L., Liu, Y., Butler, P. D., & Magid, L. J. (2007). Small Angle Neutron Scattering Studies of the Counterion Effects on the Molecular Conformation and Structure of Charged G4 PAMAM Dendrimers in Aqueous Solutions. *Macromolecules 40,* 5887–5898.
65. Nisato, G., Ivkov, R., & Amis, E. J. (2000). Size Invariance of Polyelectrolyte Dendrimers. *Macromolecules 33*, 4172–4176.
66. Gorman, C. B., & Smith, J. C. (2000). Effect of repeat unit flexibility on dendrimer conformation as studied by atomistic molecular dynamics simulations. *Polymer 41*, 675–683.
67. Blaak, R., Lehmann, S., & Likos, C. N. (2008). Charge-induced conformational changes of dendrimers. *Macromolecules 41*, 4452–4458.
68. Gitsov, I., & Fre, J. M. J. (1996). Stimuli-Responsive Hybrid Macromolecules: Novel Amphiphilic Star Copolymers With Dendritic Groups at the Periphery. *J. Am. Chem. Soc. 118*, 3785–3786.
69. Li, T., Hong, K., Porcar, L., Verduzco, R., Butler, P. D., Smith, G. S., Liu, Y., & Chen, W.-R. (2008). Assess the Intramolecular Cavity of a PAMAM Dendrimer in Aqueous Solution by Small-Angle Neutron Scattering. *Macromolecules 41*, 8916–8920.
70. Huißmann, S., Likos, C. N., & Blaak, R. (2010). Conformations of high-generation dendritic polyelectrolytes. *J. Mater. Chem. 20,* 10486–10494.
71. Maiti, P. K., & Goddard, W. A. (2006). Solvent quality changes the structure of G8 PAMAM dendrimer, a disagreement with some experimental interpretations. *J. Phys. Chem. B 110*, 25628–25632.
72. Tian, W., & Ma, Y. (2013). Theoretical and computational studies of dendrimers as delivery vectors. *Chem. Soc. Rev. 42*, 705–727.
73. Zhang, Y., Thomas, T. P., Lee, K.-H., Li, M., Zong, H., Desai, A. M., Kotlyar, A., Huang, B., Holl, M. M. B., & Baker, J. R. (2011). Polyvalent saccharide-functionalized generation 3 poly(amidoamine) dendrimer-methotrexate conjugate as a potential anticancer agent. *Bioorg. Med. Chem. 19,* 2557–2564.

74. Albertazzi, L., Brondi, M., Pavan, G. M., Sato, S. S., Signore, G., Storti, B., Ratto, G. M., & Beltram, F. (2011). Dendrimer-based fluorescent indicators: In vitro and in vivo applications. *PLoS One 6*, e28450.
75. Alcala, M.A; Kwan, S. Y., Shade, C. M., Lang, M., Uh, H., Wang, M., Weber, S. G., Bartlett, D. L., Petoud, S., & Lee, Y. J. (2011). Luminescence targeting and imaging using a nanoscale generation 3 dendrimer in an in vivo colorectal metastatic rat model. *Nanomedicine 7,* 249–258.
76. Lee, H., & Larson, R. G. (2009). A molecular dynamics study of the structure and inter-particle interactions of polyethylene glycol-conjugated PAMAM dendrimers. *J. Phys. Chem. B 113*, 13202–13207.
77. Mecke, A., Uppuluri, S., Sassanella, T. M., Lee, D.-K., Ramamoorthy, A., Baker, J. R., Orr, B. G., & Banaszak Holl, M. M. (2004). Direct observation of lipid bilayer disruption by poly(amidoamine) dendrimers. *Chem. Phys. Lipids 132*, 3–14.
78. Mecke, A., Majoros, I. J., Patri, A. K., Baker, J. R., Holl, M. M. B., & Orr, B. G. (2005). Lipid bilayer disruption by polycationic polymers: The roles of size and chemical functional group. *Langmuir 21*, 10348–10354.
79. Gardikis, K., Hatziantoniou, S., Viras, K., Wagner, M., & Demetzos, C. (2006). A DSC and Raman spectroscopy study on the effect of PAMAM dendrimer on DPPC model lipid membranes. *Int. J. Pharm. 318*, 118–123.
80. Ionov, M., Gardikis, K., Wróbel, D., Hatziantoniou, S., Mourelatou, H., Majoral, J.-P., Klajnert, B., Bryszewska, M., & Demetzos, C. (2011). Interaction of cationic phosphorus dendrimers (CPD) with charged and neutral lipid membranes. *Colloids Surf. B. Biointerfaces 82,* 8–12.
81. El-Sayed, M. E. H., Ghandehari, H., Ginski, M., & Rhodes, C. A. (2003). Influence of Surface Chemistry of Poly (Amidoamine) Dendrimers on Caco-2 Cell Monolayers. *J. Bioact. Compat. Polym. 18*, 7–22.
82. Hong, S., Leroueil, P. R., Janus, E. K., Peters, J. L., Kober, M.-M., Islam, M. T., Orr, B. G., Baker, J. R., & Banaszak Holl, M. M. (2006). Interaction of polycationic polymers with supported lipid bilayers and cells: Nanoscale hole formation and enhanced membrane permeability. *Bioconjug. Chem. 17*, 728–734.
83. Kitchens, K. M., El-Sayed, M. E. H., & Ghandehari, H. (2005). Transepithelial and endothelial transport of poly (amidoamine) dendrimers. *Adv. Drug Deliv. Rev. 57,* 2163–2176.
84. Calabretta, M. K., Kumar, A., McDermott, A. M., & Cai, C. (2007). Antibacterial activities of poly(amidoamine) dendrimers terminated with amino and poly(ethylene glycol) groups. *Biomacromolecules 8,* 1807–1811.
85. Ainalem, M.-L., Campbell, R. A., Khalid, S., Gillams, R. J., Rennie, A. R., & Nylander, T. (2010). On the Ability of PAMAM Dendrimers and Dendrimer/DNA Aggregates To Penetrate POPC Model Biomembranes. *J. Phys. Chem. B 114*, 7229–7244.
86. Ting, C. L., & Wang, Z.-G. (2011). Interactions of a charged nanoparticle with a lipid membrane: Implications for gene delivery. *Biophys. J. 100,* 1288–1297.
87. Mecke, A., Lee, D.-K., Ramamoorthy, A., Orr, B. G., & Holl, M. M. B. (2005). Synthetic and natural polycationic polymer nanoparticles interact selectively with fluid-phase domains of DMPC lipid bilayers. *Langmuir 21,* 8588–8590.

88. Tian, W., & Ma, Y. (2012). pH-responsive dendrimers interacting with lipid membranes. *Soft Matter 8*, 2627–2632.
89. Tu, C., Chen, K., Tian, W., & Ma, Y. (2013). Computational Investigations of a Peptide- Modified Dendrimer Interacting with Lipid Membranes. *Macromol. Rapid Commun. 34*, 1237–1242.
90. Xie, L., Tian, W., & Ma, Y. (2013). Computer simulations of the interactions of high-generation polyamidoamine dendrimers with electronegative membranes. *Soft Matter 9,* 9319–9325.
91. Milhem, O. M., Myles, C., McKeown, N. B., Attwood, D., & D'Emanuele, A. (2000). Polyamidoamine Starburst dendrimers as solubility enhancers. *Int. J. Pharm. 197*, 239–241.
92. Kolhe, P. (2003). Drug complexation, in vitro release and cellular entry of dendrimers and hyperbranched polymers. *Int. J. Pharm. 259*, 143–160.
93. Avila-Salas, F., Sandoval, C., Caballero, J., Guiñez-Molinos, S., Santos, L. S., Cachau, R. E., & González-Nilo, F. D. (2012). Study of Interaction Energies between the PAMAM Dendrimer and Nonsteroidal Anti-Inflammatory Drug Using a Distributed Computational Strategy and Experimental Analysis by ESI-MS/MS. *J. Phys. Chem. B 116*, 2031–2039.
94. Radhika, R., Rohith, V., Anil Kumar, N. C., Varun Gopal, K., Krishnan Namboori, P. K., & Deepak, O. M. (2010). Insilico Analysis of Nano Polyamidoamine (PAMAM) Dendrimers for Cancer Drug Delivery. *Int. J. Recent Trends Eng. Technol. 4*, 142–144.
95. Satish Kumar, M. V., & Maiti, P. K. (2012). Structure of DNA-functionalized dendrimer nanoparticles. *Soft Matter 8,* 1893–1900.
96. Nandy, B., & Maiti, P. K. (2011). DNA compaction by a dendrimer. *J. Phys. Chem. B 115,* 217–230.
97. Merkel, O. M., Zheng, M., Mintzer, M. A., Pavan, G. M., Librizzi, D., Maly, M., Höffken, H., Danani, A., Simanek, E. E., & Kissel, T. (2011). Molecular modeling and in vivo imaging can identify successful flexible triazine dendrimer-based siRNA delivery systems. *J. Control. Release 153,* 23–33.
98. Schneider, C. P., Shukla, D., & Trout, B. L. (2011). Effects of solute-solute interactions on protein stability studied using various counterions and dendrimers. *PLoS One 6,* e27665.
99. Shaunak, S., Thomas, S., Gianasi, E., Godwin, A., Jones, E., Teo, I., Mireskandari, K., Luthert, P., Duncan, R., Patterson, S., et al. (2004). Polyvalent dendrimer glucosamine conjugates prevent scar tissue formation. *Nature 22,* 977–984.
100. Teo, I. Toms, S. M., Marteyn, B., Barata, T. S., Simpson, P., Johnston, K. A., Schnupf, P., Puhar, A., Bell, T., Tang, C., et al. (2012). Preventing acute gut wall damage in infectious diarrhoeas with glycosylated dendrimers. *EMBO Mol. Med. 4*, 866–881.
101. Isea, R., Hoebeke, J., & Mayo-García, R. (2013). Designing a peptide-dendrimer for use as a synthetic vaccine against *Plasmodium falciparum. Am. J. Bioinforma. Comput. Biol. 1,* 1–8.
102. Giri, J., Diallo, M. S., Simpson, A. J., Liu, Y., Goddard, W. A., Kumar, R., & Woods, G. C. (2011). Interactions of poly(amidoamine) dendrimers with human serum albumin: Binding constants and mechanisms. *ACS Nano 5,* 3456–3468.

CHAPTER 5

DENDRIMER-DRUG INTERACTIONS

ZAHOOR AHMAD PARRY, PhD, and RAJESH PANDEY, MD

CONTENTS

5.1 INTRODUCTION

Dendrimers, are extensively branched molecules having a specific size/ shape, are a special category of nano-drug carriers. The presence of a hydrophobic core and a hydrophilic periphery imparts micelle-like performance characteristics besides the drug loading attributes, in aqueous solution [1]. The drug payloads may be either encapsulated within the dendrimer architecture via the formation of non-covalent complexes or attached to the dendrimer surface through covalent conjugation, enabling dendrimers to incorporate a lower quantity of drugs than other drug carriers [2]. On the other hand, covalently synthesized dendrimeric macromolecules have the merit of better and specific control over drug release and can be designed to limit the release of drug in the systemic circulation although ensuring to trigger the release at the target, i.e., under tumor-specific conditions [3]. Dendrimers have clearly been shown to enhance the delivery of doxorubicin and other anticancer drugs to solid tumors besides minimizing their accumulation in non-tumor tissues [4]. Dendrimers are also worthy as a tool in solubilizing sparingly soluble drugs. Polymeric dendrimers such as poly(amidoamine) dendrimers have been successfully applied to flurbiprofen, piroxicam and methotrexate for solubilization vis-a-vis targeted delivery [5–7]. Especially, lactoferrin-linked dendrimeric nano-composites demonstrated an enhanced systemic residence time and high pulmonary delivery, with a potential for reduced dosing frequency and minimal absolute dose [7]. Other examples of target selectivity include the conjugation of targeting ligands, e.g., epidermal growth factor, a rginyl glycylaspartic acid (RGD) peptides, vascular endothelial growth factor (VEGF), folate and monoclonal antibodies, onto the dendrimer surface, highlighting certain merits/demerits of these systems (Table 5.1) [8]. However, in spite of these attractive prospects, most dendrimers exhibit toxic including hemolytic activity due to their cationic surface [9]. Conceivably, anionic dendrimers as well as dendrimers suitably modified to mask these peripheral cationic groups may exhibit reduced hemolytic activity. Therefore, surface engineering of dendrimers continues to be a hot area of research and should lead to improvement of the pharmacokinetics and safety profile, in the context of their potentially wide biomedical applications.

TABLE 5.1 Merits/Demerits of Dendrimer-Based Biopharmaceuticals

Merits
• Enhanced solubility.
• High membrane permeability.
• Controlled drug release.
• Specificity.
• Low immunogenicity.
Demerits
• Suitable route of administration
• Toxicity issues

5.2 ENGINEERED NANOPARTICLES

Particle size reduction approach is extensively used to enhance the dissolution rate. This is because the rate of dissolution of a drug proportionally increases as the surface area of drug particles increases [10]. As per the 'Prandtl boundary layer equation,' the decrease in diffusion layer thickness achieved by reducing the particle size results in accelerated dissolution. Thus, based on this guideline, drug nanocrystal technology has been in the spotlight and the methodologies that have been developed to produce drug nanosuspensions chiefly include:

1. 'Bottom-up' (controlled precipitation) type.
2. 'Top-down' (wet-milling with beads, and homogenization at high-pressure) type [11, 12].

In both the bottom-up as well as the top-down approaches, the hydrophilic polymer and/or the surfactant are employed for stabilizing a nanosuspension. After a drying process (e.g., spray drying, lyophilization), the drug nanoparticles are dispersed into inert carriers. The solidified nanocrystal formulations that result are now aptly defined as 'nanocrystalline solid dispersions'.

There are many studies demonstrating the improved oral bioavailability, pharmacokinetics and pharmacodynamics of drug delivery systems obtained via nanotechnology [13]. For instance, with the nanosized formulation approach, the C_{max} (peak plasma concentration) and bioavailability were enhanced up to many folds when compared with the conventional formulations. An interesting observation is that the neutral or acidic drugs, e.g., tranilast, danazol, cilostazol, and curcumin have demonstrated better improvements in pharmacokinetics than have the basic drugs, with nanocrystal technology [14–17]. Recently, with the aid of SoluMatrix™ technology developed by iCeutica (Philadelphia, USA), a low-dose diclofenac submicron particle was developed in capsule form. In 2013, the US FDA approved this nanodrug for the therapy of acute pain of mild to moderate intensity in adults [18]. During the Phase I study, 35 mg of the oral diclofenac nanoformulations exhibited faster absorption and similar C_{max} when compared with 50 mg of diclofenac in healthy subjects. It also provided effective analgesia as observed in the Phase III study, in patients suffering from acute pain [9]. This SoluMatrix™ technology was subsequently also applied to indomethacin wherein a Phase I study documented that the oral indomethacin nanoformulations showed a faster T_{max} (time to C_{max}) of 1.1 hours in comparison with 2 hours for unformulated indomethacin, possibly resulting in faster onset of action [20]. The C_{max} for nanoformulated 40 mg indomethacin was slightly more compared with the standard 50 mg oral indomethacin in healthy subjects. Thus, it is reasonable to conclude that nanoformulated systems may help to reduce the dose of drugs, thereby improving their tolerability and safety while maintaining their efficacy. In general, neutral/acidic drugs would be poorly soluble in gastric juice and the improved dissolution pattern of these drugs under at acidic pH via nanotechnology can lead to significant enhancement in the oral bioavailability. Nanoparticles are known to exhibit mucoadhesion for the intestinal mucosa, and this property also contributes to the enhancement of oral bioavailability. The transcellular and paracellular uptake of nanoparticles through the intestinal barrier is an added factor that helps to improve the oral absorption, and this uptake is influenced by several factors, e.g., particle size and surface charge [21]. Therefore, the mechanisms for improved oral absorption may be summarized as:

1. Improved dissolution behavior.
2. Bioadhesion to the intestinal epithelium.
3. Transcellular and paracellular uptake.

Besides the improved dissolution and bioavailability, nanocrystals have also provided a very important additional pharmaceutical benefit, i.e., enhanced patient compliance because of the reduction in dosing frequency [22]. Research has revealed that nanocrystal formulations can be applied to dermal, ocular as well as pulmonary drug delivery. For instance, in a rat model of airway inflammation, inhalable tranilast as nanocrystalline solid dispersion has been prepared by the wet-milling technology, and it exhibited good inhalation characteristics and enhanced anti-inflammatory effects when compared with crystalline tranilast having a larger particle size [23].

5.3 ENCAPSULATION AND CONTROLLED RELEASE

Dendrimers have been widely used as vehicles for drug as well as gene delivery because of their property to entrap these compounds in their interior or on their surface [24–29]. Layer-by-layer (LbL)-deposited thin films or microcapsules have also been evaluated as scaffolds for controlled drug release [30–34]. In this context, the dendrimer-containing LbL films are likely to have promising applications in drug/gene delivery, targeting and controlled release. In addition, the dendrimer-containing LbL films have the potential for providing multiple binding sites for the drugs in the polymer interior, on the surface, and in the complete film.

LbL assemblies comprised of PAMAM dendrimer and PSS were made to deposit on the surface of a flat substrate as well as on polystyrene beads for studying the loading and release of CF (carboxylated fluorescein), a dye [35]. This model dye was entrapped in the PAMAM/PSS films following exposure of the films to the CF solution. Tertiary as well as primary amine residues of PAMAM most probably provide the linking sites to the anionic CF. The bound CF was subsequently released from LbL films into NaCl solution (0.154 M, pH 6.5) as per Fickian-type kinetics. The release rate was high and >75% of bound CF was released during the first 60 min. Other researchers have reported the preparation of dendrimer/liposome LbL films for loading and release of ibuprofen [36, 37]. It is well appreciated that stimulus-sensitive materials are critical for the development of drug delivery systems. To achieve this goal, LbL films sensitive to sugars, salts, pH, temperature and electric signals have been studied [38–41]. Hydrogen-bonding LbL films are good examples of pH-sensitive films since poly(carboxylic acid) (which is a common component of LbL films) dissociates at neutral or basic pH, thereby resulting in the breakage of hydrogen bonds. By employing this strategy, researchers have shown the pH-induced release of film components from PAMAM-COOH/PVP LbL [42]. These PAMAM-COOH/PVP films were found to be stable at acidic and neutral pH but unstable at pH 12 and 13. Further, PAMAM-COOH/PMA and PAMAM-COOH/PAA films were observed to be more sensitive to pH changes since these films decomposed either at pH 5.5 and 5.0 or higher respectively [43, 44]. Other model dyes, for instance Rose Bengal and sulfonated tetraphenylporphyrin, were also reported to be loaded and released from PAMAM-COOH/PMA films as a response to changes in pH [43].

Recently, attention has focused on LbL microcapsules prepared via the alternating laying down of polymers on the surface of colloid particles used as template, subsequently followed by dissolution of the template material (Figure 5.1) [30–34]. Such LbL microcapsules bearing dendrimer shells are interesting since they provide two distinct linking sites in the capsule interior and on the shell. Thus, PAMAM/PSS film-based microcapsules were studied for the loading and release of DOX (doxorubicin, an anticancer drug) [45]. A sustained release of DOX from the microcapsules was noted in 0.154 M NaCl solution, lasting several hours. By using Au nanoparticle-encapsulating PAMAM, the stability of PAMAM/PSS microcapsules was enhanced [46]. LbL microcapsules bearing phosphorus dendrimers linked to linear polymers or DNA have been studied with respect to their mechanical properties. The phosphorus dendrimer microcapsules were softer as compared to microcapsules that were assembled from linear polyelectrolytes although the microcapsules were hardened by treatment with organic solvent [47–49].

Protein-loaded microcapsules are potentially useful for the selective/controlled encapsulation of various biomolecules [50, 51]. AgSD (silver sulfadiazine, an antibiotic with poor water solubility) was coupled with LbL films made up of oppositely charged PAMAM dendrimers for improving the aqueous solubility and hence the antibiotic activity [52]. In this preparation, AgSD microparticles were directly coated with the dendrimer LbL films. As a result, the loading of drug into the microcapsules was high, in sharp contrast to the loading of drug into previously synthesized microcapsules. The topical formulations, i.e., PAMAM LbL film coated AgSD nanoparticles demonstrated significantly higher antibacterial activity compared with the formulation lacking the PAMAM coating. It is emphasized that the dendrimer itself has antimicrobial including antifungal activities [53–55].

It is also possible to encapsulate the PAMAM dendrimers in microcapsules prepared by using poly(allylamine) or poly(vinyl sulfate) as the shell materials. These PAMAM-containing microcapsules had been fur-

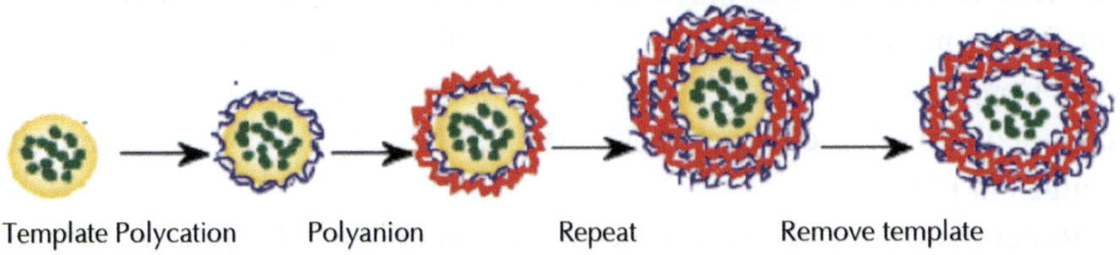

FIGURE 5.1 Template-based LbL synthesis. (From Sato, K.; Takahashi, S.; Anzai, J. Layer-by-layer thin films and microcapsules for biosensors and controlled release. Anal. Sci. 2012, 28, 929–938 (ISSN: 1348-2246). Copyright (2011) The Japan Society for Analytical Sciences.)

ther loaded with ANS (fluorescent 1-anilinonaphthalene-8-sulfonic acid, a model drug) by linking ANS to the dendrimers. It was noted that the rate of uptake of ANS was largely determined by the rate of transport of ANS across the capsule shell. On the other hand, the dissociation of ANS from the PAMAM turned out to be the rate-determining step for the release of ANS out of the capsule [56].

Thus, dendrimer-containing LbL films as well as microcapsules have found diverse applications in biotechnology, construction of biosensors, drug delivery systems and nanomedicine as a whole. The reasons for the success of dendrimer-LbL concept are multifactorial as already discussed (Table 5.2).

5.4 INTERACTIONS BETWEEN DENDRIMERS AND DRUG MOLECULES

The functionality of dendrimer end groups can be modified for obtaining molecules having novel biological properties, e.g., cooperative receptor-ligand interactions that would help in the interaction of dendrimers with poorly soluble drugs. Dendrimers are capable of enhancing the cellular uptake, optimizing biodistribution, improving bioavailability and therapeutic efficacy, and reducing the systemic toxicity, clearance and rate of degradation of drugs [57]. There are two techniques for dendrimer-based drug delivery: (i) the lipophilic drugs can be entrapped

TABLE 5.2 Key Factors Behind the Success of Dendrimer-LbL Concept

- Successful assembly of dendrimer-LbL films/microcapsules via electrostatic bonding, hydrogen bonding, covalent bonding, and biological affinity.
- Facile preparation of chemically modified dendrimers.
- Surface reactive moieties of dendrimers (e.g., primary amine and carboxylic acid residues) may be covalently/non-covalently altered with desired functional groups.
- Possibility of accommodating small molecules/nanoparticles in the interior of dendrimers.
- Acceptable biocompatibility of modified dendrimers.
- Feasibility for targeting and imaging applications, e.g., Fe_3O_4 nanoparticles coated with dendrimer LbL film.

within the hydrophobic cavity of dendrimer to make them hydrophilic; and (ii) drugs may be covalently linked onto the dendrimer surface. Encapsulation actually traps the drug inside the dendrimer by exploiting the interaction between drug and dendrimer or simply the steric bulk of the dendrimer exterior. Further, the nature of drug encapsulation may be a simple physical entrapment or may involve specific non-bonding interactions within the dendrimer. However, it should be appreciated that the drug is linked to the exterior of the dendrimer in case of dendrimer-drug conjugates. The conjugates are generally prodrugs (which are either inactive as such or weakly active). The covalent conjugation of drugs was largely employed for targeting and achieving higher drug payload, while the noncovalent interactions resulted in greater solubility of otherwise insoluble drugs [58, 59]. Obviously, the unique architectural pattern of dendrimers also makes them ideal for the preparation of cross-linked covalent gels as well as for the self-assembled non-covalent gels [60].

Dendrimer-drug interactions are known to be affected by several factors such as the structure, generation, surface engineering, and concentration of the dendrimers. For example, PPI and PAMAM have a slightly different framework (internal branches) and this difference in makes PPI dendrimers more hydrophobic as compared to PAMAM dendrimers thereby resulting in variable solubility [61]. Further, the modification of dendrimer surface may enhance the therapeutic efficacy of drugs as far as better targeting and reduction in toxicity are concerned.

5.5 ENCAPSULATION OF DRUGS WITHIN DENDRITIC STRUCTURE

The acid-base reaction between dendrimers and the molecules of interest such as drugs (often termed 'guest molecules') with Coulomb attractions pull the guest molecules inside the dendrimer. On the other hand, the hydrogen bonding holds them together. The first report on the encapsulation of a dye inside a dendrimer appeared in 1994, the so-called 'dendritic box' [62]. The guest molecules can be entrapped in the internal cavities of dendrimers during the synthetic process and this is aided by the shell preventing diffusion, even following prolonged harsh treatment such as heating, sonication, and solvent extraction [63, 64]. Subsequent to the successful encapsulation of dyes in dendrimers, the anticancer drug encapsulation became the focus of research, as exemplified by the entrapment of the anticancer drugs methotrexate and doxorubicin. G3 and G4 PAMAM dendrimers with poly(ethylene glycol) mono-methyl ether (M-PEG) grafts were employed by researchers [65]. Further, the same group of workers also attached methotrexate and folic acid to the exterior of dendrimers and targeted tumor cells with the drug-dendrimer conjugates [66].

Dendrimers with polar shell and apolar core have been referred to as 'unimolecular or unilamellar micelles'. The dendritic structure does not depend on the dendrimer concentration unlike the conventional micelles [67]. However, this approach faces a drawback that it is not easy to control the release of entrapped molecules from the dendrimer core. PEG has been used for modifying dendrimers by linking PEG with the surface of dendrimers and thereby forming a unimolecular micelle (obviously by providing a hydrophilic shell surrounding the dendrimer core). PEGylated dendrimers are of special interest in drug delivery systems because of several attributes (Table 5.3) [68, 69].

Researchers have also synthesized 'cored dendrimers' with modified dendrimeric architecture for drug encapsulation. Following the synthesis of dendrimer, the core was removed by cleaving the ester bonds, while the remaining structure was the same as a result of robust ether linkages [70, 71]. Further, 'dendrimeric block copolymers' have been synthesized bearing linear hydrophilic blocks as well as a hydrophobic dendritic block. The ability of these block copolymers to complex poorly aqueous soluble molecules has been investigated. The synthesis of a series of triblock copolymers, i.e.,

TABLE 5.3 Special Attributes of PEGylated Dendrimers

- Ease of preparation.
- East availability of PEG.
- Low cost of PEG.
- Biocompatibility
- Higher water solubility.
- Ability to modify the biodistribution of carrier systems.

PAMAM-block-PEG-block-PAMAM (G1–G5) was studied as potential carriers for gene delivery [72].

Though dendrimer-mediated complexation has several merits in terms of stability, controlled release, high drug loading, and reduced toxicity of entrapped drugs, the noncovalent complex formation may sometimes lead to lower drug entrapment and compromised stability when compared with covalent conjugation [73].

5.6 DENDRIMER-DRUG CONJUGATION

The outer surfaces of dendrimers are of unique interest and have been thoroughly investigated as potential sites of interaction with drugs of interest for increasing the loading capacity. The number of surface groups available for drug interactions increases by nearly two-folds per higher generation of dendrimer. Drugs may be conjugated to dendrimeric systems via ester, amide, or other suitable linkages depending on the dendrimer surface. Once inside the cell, the linkage should be amenable to hydrolysis by endosomal/lysosomal enzymes [57–59]. There are several ionizable groups on the dendrimer surface and these are the sites where ionizable drugs may link via electrostatic interactions. Several linkages are possible (Table 5.4) [74].

A lot of research data on drug loaded-dendrimers shows that the release of the free drug could be enhanced by a suitably chosen linker. The linker/spacer length and flexibility are two crucial factors. Further, some of the linkers may be pH-sensitive and have been documented to boost the intracellular release of free drug [75]. The comparison of covalently tagged drug and noncovalent inclusion complex with respect to

TABLE 5.4 Types of Drug-Dendrimer Surface Conjugates

- Amides.
- Esters.
- Disulfides.
- Hydrazones.
- Thiol-maleimide.
- Sulfinyl.

the release kinetics and efficacy, using G5 PAMAM dendrimers for targeting methotrexate, revealed that the inclusion complex released the drug immediately with the drug being active *in vitro*. On the other hand, the covalently conjugated drug was better suited for targeted drug delivery [76]. A better control over drug release could be achieved by the covalent drug tagging employing biodegradable linkages compared to electrostatic drug-dendrimer complexes. The major demerit, however, of the conjugation process is the possibility of drug release being less active (too sluggish) and consequently losing the potential to be efficacious *in vivo* [77].

5.7 DENDRITIC GELS

The pharmaceutical applications of hydrogels have gained attention because of their structural properties, good drug loading capacity, and capability to ensure controlled drug release. Hydrogels possess high water absorbing capacity [78]. The '*in situ* forming gels' have been evaluated for a wide range of applications such as oral, ocular, nasal, rectal, vaginal, and injectable [79, 80]. Dendrimers could be prepared with carefully controlled and customized molecular sizes possessing favorable properties between those of traditional gel polymers and organic compounds having low molecular weight employed in the 'self assembled supramolecular gels' [60]. A so-called 'polymer network' is generally obtained by using a cross-linker during the stage of polymerization. The synthesis of hydrogels remains a function of the multivalent cross-linking nature of dendrimers [81]. It has been demonstrated that the dendrimeric branching has a key role in controlling the self-assembly. Further, the repeated use of its critical structural motifs could result in multiple interactions between the branched units and thus lend strength to the noncovalent interactions which are responsible for the self-assembly process [82].

With this background regarding dendrimer-drug interactions, we now proceed to discuss specific instances of loading of molecules of key clinical interest (drugs, genes) into dendrimers.

KEYWORDS

- **biopharmaceuticals**
- **dendrimer interactions**
- **dendritic structure**
- **nanocrystals**
- **nanoparticles engineering**
- **sustained release**

REFERENCES

1. Liu, M., & Fréchet, J. M. J. (1999). Designing dendrimers for drug delivery. *Pharm Sci Technolo Today. 2*(10), 393–401.
2. Honda, M., Asai, T., Oku, N., Araki, Y., Tanaka, M., & Ebihara, N. (2013). Liposomes and nanotechnology in drug development: focus on ocular targets. *Int J Nanomedicine. 8*, 495–503.
3. Kaminskas, L. M., McLeod, V. M., Kelly, B. D., et al. (2012). A comparison of changes to doxorubicin pharmacokinetics, antitumor activity, and toxicity mediated by PEGylated dendrimer and PEGylated liposome drug delivery systems. *Nanomedicine. 8*(1), 103–111.
4. Maeda, H., Sawa, T., & Konno, T. (2001). Mechanism of tumor-targeted delivery of macromolecular drugs, including the EPR effect in solid tumor and clinical overview of the prototype polymeric drug SMANCS. *J Control Release. 74*(1–3), 47–61.
5. Asthana, A., Chauhan, A. S., Diwan, P. V., & Jain, N. K. (2005). Poly(amidoamine) (PAMAM) dendritic nanostructures for controlled site-specific delivery of acidic anti-inflammatory active ingredient. *AAPS Pharm Sci Tech. 6*(3), E536–E542.
6. Prajapati, R. N., Tekade, R. K., Gupta, U., Gajbhiye, V., & Jain, N. K. (2009). Dendimer-mediated solubilization, formulation development and in vitro-in vivo assessment of piroxicam. *Mol Pharm. 6*(3), 940–950.
7. Kurmi, B. D., Gajbhiye, V., Kayat, J., & Jain, N. K. (2011). Lactoferrin-conjugated dendritic nanoconstructs for lung targeting of methotrexate. *J Pharm Sci. 100*(6), 2311–2320.
8. Kaminskas, L. M., & Porter, C. J. (2011). Targeting the lymphatics using dendritic polymers (dendrimers). *Adv Drug Deliv Rev. 63*(10–11), 890–900.
9. Ziemba, B., Matuszko, G., Bryszewska, M., & Klajnert, B. (2012). Influence of dendrimers on red blood cells. *Cell Mol Biol Lett. 17*(1), 21–35.
10. Kawabata, Y., Wada, K., Nakatani, M., Yamada, S., & Onoue, S. (2011). Formulation design for poorly water-soluble drugs based on biopharmaceutics classification system: basic approaches and practical applications. *Int J Pharm. 420*(1), 1–10.

11. Gao, L., Liu, G., Ma, J., Wang, X., Zhou, L., & Li, X. (2012). Drug nanocrystals: In vivo performances. *J Control Release. 160*(3), 418–430.
12. Chan, H. K., & Kwok, P. C. (2011). Production methods for nanodrug particles using the bottom-up approach. *Adv Drug Deliv Rev. 63*(6), 406–416.
13. Onoue, S., Yamada, S., & Chan, H.-K. (2014). Nanodrugs: pharmacokinetics and safety. *Int J Nanomed 9*, 1025–1037.
14. Kawabata, Y., Yamamoto, K., Debari, K., Onoue, S., & Yamada, S. (2010). Novel crystalline solid dispersion of tranilast with high photostability and improved oral bioavailability. *Eur J Pharm Sci. 39*(4), 256–262.
15. Wu, C. Y., & Benet, L. Z. (2005). Predicting drug disposition via application of BCS: transport/absorption/elimination interplay and development of a biopharmaceutics drug disposition classification system. *Pharm Res. 22*(1), 11–23.
16. Jinno, J., Kamada, N., Miyake, M., et al. (2006). Effect of particle size reduction on dissolution and oral absorption of a poorly water-soluble drug, cilostazol, in beagle dogs. *J Control Release. 111*(1–2), 56–64.
17. Onoue, S., Takahashi, H., Kawabata, Y., et al. (2010). Formulation design and photochemical studies on nanocrystal solid dispersion of curcumin with improved oral bioavailability. *J Pharm Sci. 99*(4), 1871–1881.
18. Manvelian, G., Daniels, S., & Gibofsky, A. (2012). The pharmacokinetic parameters of a single dose of a novel nano-formulated, lower-dose oral diclofenac. *Postgrad Med. 124*(1), 117–123.
19. Gibofsky, A., Silberstein, S., Argoff, C., Daniels, S., Jensen, S., & Young, C. L. (2013). Lower-dose diclofenac submicron particle capsules provide early and sustained acute patient pain relief in a phase 3 study. *Postgrad Med. 125*(5), 130–138.
20. Manvelian, G., Daniels, S., & Altman, R. (2012). A phase I study evaluating the pharmacokinetic profile of a novel, proprietary, nano-formulated, lower-dose oral indomethacin. *Postgrad Med. 124*(4), 197–205.
21. Pandey, R., & Khuller, G. K., Novel delivery systems for anti-tuberculosis drugs. In: Wing, W., Yew, Iseman, M. (Eds.). Development of New Anti-tuberculosis Drugs, 2006. Nova Science Publishers, NY, Chapter XII, p. 259–275. ISBN: 1-59454-857-9.
22. Shegokar, R., & Müller, R. H. (2010). Nanocrystals: industrially feasible multifunctional—formulation technology for poorly soluble actives. *Int J Pharm. 399*(1–2), 129–139.
23. Onoue, S., Aoki, Y., Kawabata, Y., et al. (2011). Development of inhalable nanocrystalline—solid dispersion of tranilast for airway inflammatory diseases. *J Pharm Sci. 100*(2), 622–633.
24. Beezer, A. E., King, A. S. H., Martin, I. K., Mitchel, J. C., Twyman, L. J., & Wain, C. F. (2003). Dendrimers as potential drug carriers; encapsulation of acidic hydrophobes within water soluble PAMAM dendrimers. *Tetrahedron 59*, 3873–3880.
25. Morgan, E. J., Rippy, J. M., & Tucer, S. A. (2006). Spectroscopic characterization of poly(amidoamine) dendrimers as selective uptake devices: Phenol blue versus Nile red. *Appl. Spectr. 60*, 551–559.
26. Gurdag, S., Khandare, J., Stapels, S., Matherly, L. H., & Kannan, R. M. (2006). Activity of dendrimer-methotrexate conjugates on methotrexate-sensitive and resistant cell lines. *Bioconjug. Chem. 17*, 275–283.

27. Majoros, I. J., Myc, A., Thomas, T., Mehta, C. B., & Baker, J. R., Jr. (2006). PAMAM dendrimer–based multifunctional conjugate for cancer therapy: Synthesis, characterization, and functionality. *Biomacromolecules 7*, 572–579.
28. Kazzouli, S. E., Mignani, S., Bousmina, M., & Majoral, J.-P. (2012). Dendrimer therapeutics: Covalent and ionic attachments. *New J. Chem. 36*, 227–240.
29. Ainalem, M.-L., & Nylander, T. (2011). DNA condensation using cationic dendrimers-morphology and molecular structure of formed aggregates. *Soft Matt. 7*, 4577–4594.
30. De Koker, S., Hoogenboom, R., & De Geest, B. G. (2012). Polymeric multilayer capsules for drug delivery. *Chem. Soc. Rev. 41*, 2867–2884.
31. der Mercato, L. L., Rivera–Gil, P., Abbasi, A. Z., Ochs, M., Ganas, C., Zins, I., Sonnichsen, C., & Parak, W. J. (2010). LbL multilayer capsules: recent progress and future outlook for their use in life sciences. *Nanoscale 2*, 458–467.
32. Sato, K., Yoshida, K., Takahashi, S., & Anzai, J. (2011). pH- and sugar-sensitive layer-by-layer films and microcapsules for drug delivery. *Adv. Drug Deliv. Rev. 63*, 809–821.
33. Ariga, K., McShane, M., Lvov, Y. M., Ji, Q., & Hill, J. P. (2011). Layer-by-layer assembly for drug delivery and related applications. *Exp. Opin. Drug Deliv. 8*, 633–644.
34. Sato, K., Takahashi, S., & Anzai, J. (2012). Layer-by-layer thin films and microcapsules for biosensors and controlled release. *Anal. Sci. 28*, 929–938.
35. Khopade, A. J., & Caruso, F. (2002). Electrostatically assembled polyelectrolyte/dendrimer multilayer films as ultrathin nanoreservoirs. *Nano Lett. 2*, 415–418.
36. Moraes, M. L., Baptista, M. S., Itri, R., Zucolotto, V., & Oliveira, O. N., Jr. (2008). Immobilization of liposomes in nanostructured layer-by-layer films containing dendrimers. *Mater. Sci. Eng. C 28*, 467–471.
37. Geraldo, V. P. N., Moraes, M. L., Zucolotto, V., & Oliveira, O. N., Jr. (2011). Immobilization of ibuprofen-containing nanospheres in layer-by-layer films. *J. Nanosci. Nanotechnol. 11*, 1167–1174.
38. Sato, K., Imoto, Y., Sugama, J., Seki, S., Inoue, H., Odagiri, T., Hoshi, T., & Anzai, J. (2005). Sugar-induced disintegration of layer-by-layer assemblies composed of concanavalin A and glycogen. *Langmuir 21*, 797–799.
39. Gui, Z., Qian, J., An, Q., Zhao, Q., Jin, H., & Du, B. (2010). Layer-by-layer self-assembly, controllable disintegration of polycarboxybetaine multilayers and preparation of free-standing films at physiological conditions. *J. Mater. Chem. 20*, 1467–1474.
40. Nolan, C. M., Serpe, M. J., & Lyon, L. A. (2004). Thermally modulated insulin release from microgel thin films. *Biomacromolecules 5*, 1940–1946.
41. Sato, H., Takano, Y., Sato, K., & Anzai, J. (2009). Electrochemically controlled release of αβγδ-tetrakis (4-N-methylpyridyl) porphine from layer-by-layer thin films. *J. Colloid Interface Sci. 333*, 141–144.
42. Fu, Y., Chen, H., Bai, S., Huo, F., Wang, Z., & Zhang, X. (2003). Base-induced release of molecules from hydrogen bonding directed layer-by-layer film. *Chin. J. Polym. Sci. 21*, 499–503.

43. Tomita, S., Sato, K., & Anzai, J. (2008). Layer-by-layer assembled thin films composed of carboxyl-terminated poly(amidoamine) dendriemr as a pH-sensitive nanodevice. *J. Colloid Interface Sci. 326,* 35–40.
44. Tomita, S., Sato, K., & Anzai, J. (2009). pH-Stability of layer-by-layer thin films composed of carboxyl-terminated poly(amidoamine) dendrimer and poly(acrylic acid). *Kobunshi Ronbunshu 66,* 75–78.
45. Khopade, A. J., & Caruso, F. (2002). Stepwise self-assembled poly(amidoamine) dendrimer and poly(styrenesulfonate) microcapsules as sustained delivery vehicles. *Biomacromolecules 3,* 1154–1162.
46. Levedeva, O. V., Kim, B.-S., Gröhn, F., & Vinogradova, O. I. (2007). Dendrimer-encapsulated gold nanoparticles as building blocks for multilayer microshells. *Polymer 48,* 5024–5029.
47. Kim, B.-S., Lebedeva, O. V., Kim, D. H., Caminade, A.-M., Majoral, J.-P., Knoll, W., & Vinogradova, O. I. (2005). Assembly and mechanical properties of phosphorus dendrimer/polyelectrolyte multilayer microcapsules. *Langmuir 21,* 7200–7206.
48. Kim, B.-S., Lebedeva, O. V., Koynov, K., Gong, H., Caminade, A.-M., Majoral, J.-P., & Vinogradova, O. I. (2006). Effect of dendrimer generation on the assembly and mechanical properties of DNA/phosphorus dendrimer multilayer microcapsules. *Macromolecules 39,* 5479–5483.
49. Kim, B.-S., Lebedeva, O. V., Park, M.-K., Knoll, W., Caminade, A.-M., Majoral, J.-P., & Vinogradova, O. I. (2010). THF-induced stiffening of polyelectrolyte/phosphorus dendrimer multilayer microcapsules. *Polymer 51,* 4525–4529.
50. Endo, Y., Sato, K., & Anzai, J. (2011). Preparation of avidin-containing polyelectrolyte microcapsules and their uptake and release properties. *Polym. Bull. 66,* 711–720.
51. Endo, Y., Sato, K., Yoshida, K., & Anzai, J. (2011). Avidin/PSS membrane microcapsules with biotin-binding activity. *J. Colloid Interface Sci. 360,* 519–524.
52. Strydom, S. J., Rose, W. E., Otto, D. P., Liebenberg, W., & de Villiers, M. M. (2013). Poly(amidoamine) dendrimer-mediated synthesis and stabilization of silver sulfonamide nanoparticles with increased antibacterial activity. *Nanomedicine 9,* 85–93.
53. Calabretta, M. K., Kumar, A., McDermott, A. M., & Cai, C. (2007). Antibacterial activities of poly(amidoamine) dendrimers terminated with amono and poly(ethylene glycol) groups. *Biomacromolecules 8,* 1807–1811.
54. Polcyn, P., Jurczak, M., Rajnisz, A., Solecka, J., & Urbanczyk-Lipkowska, Z. (2009). Design of antimicrobially active small amphiphilic peptide dendrimers. *Molecules 14,* 3881–3905.
55. Janiszewska, J., Sowińska, M., Rajinisz, A., Solecka, J., Łącka, I., Milewski, S., & Urbanczyk-Lipkowska, Z. (2012). Novel dendrimeric lipopeptides with antifungal activity. *Bioorg. Med. Chem. Lett. 22,* 1388–1393.
56. Tomita, S., Sato, K., & Anzai, J. (2009).Preparation of dendrimer-loaded microcapsules by a layer-by-layer deposition of polyelectrolytes. *Mater. Sci. Eng. C 29,* 2024–2028.
57. D'Emanuele, A., & Attwood, D. (2005). "Dendrimer-drug interac-tions, " *Advanced Drug Delivery Reviews, 57*(15), pp. 2147–2162.
58. Gillies, E. R. & Frechet, J. M. (2005). "Dendrimers and dendritic polymers in drug delivery," *Drug Discovery Today, 10*(1), pp. 35–43.

59. Jain, N. K., & Gupta, U. (2008). "Application of dendrimer-drug complexation in the enhancement of drug solubility and bioavailability," *Expert Opinion on Drug Metabolism and Toxicology, 4*(8), 1035–1052.
60. Smith, D. K. (2006). "Dendritic gels: many arms make light work," *Advanced Materials, 18*(20), 2773–2778.
61. Richter-Egger, D. L., Tesfai, A., & Tucker, S. A. (2001). "Spectroscopic investigations of poly(propyleneimine)dendrimers using the solvatochromic probe phenol blue and comparisons to poly(amidoamine) dendrimers," *Analytical Chemistry, 73*(23), 5743–5751.
62. Jansen, J. F. G. A., Debrabandervandenberg, E. M. M., & Meijer, E. W. (1994). "Encapsulation of guest molecules into a dendritic box," *Science, 266*(5188), 1226–1229.
63. Jansen, J. F. G. A., Debrabandervandenberg, E. M. M., & Meijer, E. W. (1995). "The dendritic box and bengal rose," *Polymeric Materials, 73*, 123–124.
64. Jansen, J. F. G. A., Meijer, E. W., & Debrabandervandenberg, E. M. M. (1995). "The dendritic box: shape-selective liberation of encapsulated guests," *Journal of the American Chemical Society, 117*(15), 4417–4418.
65. Kojima, C., Kono, K., Maruyama, K., & Takagishi, T. (2000). "Synthesis of polyamidoamine dendrimers having poly (ethylene glycol) grafts and their ability to encapsulate anticancer drugs," *Bioconjugate Chemistry, 11*(6), 910–917.
66. Kono, K., Liu, M., & Frechet, J. M. J. (1999). "Design of dendritic macromolecules containing folate or methotrexate residues, "*Bioconjugate Chemistry, 10*(6), 1115–1121.
67. Stevelmans, S., vanHest, J. C. M., Jansen, J., F. G. A., van Boxtel, D., A. F. J., Vanden Berg, E. M. M. D. & Meijer, E. W. (1996). "Synthesis, characterization and guest-host properties of inverted uni-molecular dendritic micelles," *Journal of the American Chemical Society, 118*(31), 7398–7399.
68. Pan, G. F., Lemmouchi, Y., Akala, E. O., & Bakare, O. (2005). "Studies on pegylated and drug-loaded pamam dendrimers," *Journal of Bioactive and Compatible Polymers, 20*(1), 113–128.
69. Greenwald, R. B., Choe, Y. H., McGuire, J., & Conover, C. D. (2003). "Effective drug delivery by pegylated drug conjugates," *Advanced Drug Delivery Reviews, 55*(2), 217–250.
70. Wendland, M. S., & Zimmerman, S. C. (1999). "Synthesis of cored dendrimers," *Journal of the American Chemical Society, 121*(6), 1389–1390.
71. Schultz, L. G., Zhao, Y., & Zimmerman, S. C. (2001). "Synthesis of cored dendrimers with internal cross-links," *Angewandte Chemie International Edition, 40*(10), 1962–1966.
72. Kim, T., Seo, H. J., Choi, J. S. et al. (2004). "Pamam-peg-pamam: novel triblock copolymer as a biocompatible and efficient gene deliv-ery carrier," *Biomacromolecules, 5*(6), 2487–2492.
73. Boas, U., & Heegaard, P. M., (2004). "Dendrimers in drug research," *Chemical Society Reviews, 33*(1), 43–63.
74. Duncan, R., & Izzo, L. (2005). "Dendrimer biocompatibility and toxicity," *Advanced Drug Delivery Reviews, 57*(15), 2215–2237.

75. S. El Kazzouli, Mignani, S., Bousmina, M., & Majoral, J. P. (2012). "Dendrimer therapeutics: covalent and ionic attachments," *New Journal of Chemistry, 36*(2), 227–240.
76. Patri, A. K., Kukowska-Latallo, J. F., & Baker, J. R. (2005). "Targeted drug delivery with dendrimers: comparison of the release kinetics of covalently conjugated drug and non-covalent drug inclusion complex," *Advanced Drug Delivery Reviews, 57(*15), 2203–2214.
77. Kaminskas, L. M., McLeod, V. M., Porter, C. J., & Boyd, B. J. (2012). "Association of chemotherapeutic drugs with dendrimer nanocarriers: an assessment of the merits of covalent conjugation compared to noncovalent encapsulation," *Molecular Pharmaceutics, 9*(3), 355–373.
78. Hoare, T. R., & Kohane, D. S. (2008). "Hydrogels in drug delivery: progress and challenges," *Polymer, 49*(8), 1993–2007.
79. Nirmal, H. B., Bakliwal, S. R., & Pawar, S. P. (2010). "In-situ gel: new trends in controlled and sustained drug delivery system," *International Journal of PharmTech Research, 2*(2), 1398–1408.
80. Navath, R. S., Menjoge, A. R., Dai, H., Romero, R., Kannan, S., & Kannan, R. M. (2011). "Injectable pamam dendrimer-peg hydrogels for the treatment of genital infections: formulation and in vitro and in vivo evaluation," *Molecular Pharmaceutics, 8*(4), 1209–1223.
81. Sontjens, S. H. M., Nettles, D. L., Carnahan, M. A., Setton, L. A., & Grinstaff, M. W. (2006). "Biodendrimer-based hydrogel scaffolds for cartilage tissue repair," *Biomacromolecules, 7*(1), 310–316.
82. Jang, W. D., & Aida, T. (2003). "Dendritic physical gels: structural parameters for gelation with peptide-core dendrimers," *Macromolecules, 36*(22), 8461–8469.

CHAPTER 6

DENDRIMERS AND DRUG DELIVERY

ZAHOOR AHMAD PARRY, PhD, and RAJESH PANDEY, MD

CONTENTS

6.1 INTRODUCTION

In 1982, dendrimers were proposed for the first time for their potential application as 'molecular containers' [1]. The dendrimer-drug (often called 'host-guest') properties of various dendrimeric polyplexes have been extensively investigated and have attained a key position in the domain of supramolecular chemistry [2]. The dendrimer-drug chemistry is based on the reaction of linking of a substrate (e.g., drug) to the carrier, i.e., dendrimer whose details were covered in Chapter 5. Notwithstanding the versatility of dendrimers as carriers for a wide category of drugs, here we restrict ourselves to certain specialized aspects only.

6.2 TRANSDERMAL DRUG DELIVERY

The clinical use of non-steroidal anti-inflammatory drugs (NSAIDs) is limited owing to a number of undesirable reactions such as gastro-intestinal

(GI) side effects (e.g., mucosal ulceration), renal side effects (e.g., tubular damage), etc., following oral administration. Transdermal drug delivery can overcome these adverse effects besides maintaining therapeutic drug levels in plasma for a longer period of time. However, transdermal delivery is challenged with poor rates of transcutaneous drug delivery owing to barrier functions of the skin. Dendrimers have been applied to transdermal drug delivery systems with success. In general, for drugs possessing hydrophobic moieties in their structure and having low aqueous solubility, dendrimers offer a good choice in the field of drug delivery systems [3].

Dendrimers are superior when compared with the other currently available drug delivery systems such as liposomal and linear polymer formulations. This is because of their defined architectural control, high drug loading, and precise as well as controlled drug loading/drug release. Needless to mention that each of these factors are critical for co-delivery. Dendrimers are especially useful for co-delivery since they can entrap poorly water soluble drugs in the hydrophobic cores and the versatile surface moieties allow the attachment of multiple drugs onto the surface [4]. Further, they may be suitably complexed with nucleic acids and employed in gene therapy (Chapter 7). This is in sharp contrast to the conventional linear polymers that have a random coil-like architecture and thus offer less structural control.

Dendrimers also possess the capability of simultaneously loading hydrophobic and hydrophilic drugs. Researchers have used this property for conjugating the hydrophobic drug paclitaxel and the hydrophilic drug alendronate with dendritic poly(ethylene glycol). A high degree of homogeneity of the fixed paclitaxel/alendronate ratio also demonstrated better targeting and enhanced activity when compared with the free drugs [5].

Achieving a high drug loading is one of the critical parameters in drug delivery and it is all the more important in co-delivery since two different drugs need to be loaded in the same system [6]. As mentioned before, dendrimers offer the potential of high drug loading due to the hydrophobic core and numerous surface groups that are available for modification. One group of researchers took advantage of these multiple surface groups for achieving both high drug loading vis-a-vis pH sensitivity. This was accomplished by conjugating the drug doxorubicin (DOX) to MPEG PAMAM via an acid labile hydrazine linkage. This conjugate could self-assemble, with DOX behaving as a hydrophobic core. Subsequently, 10-hydroxy-camptothecin (HCPT) was introduced into the hydrophobic core. It was observed that the system displayed a high drug loading, i.e., 41.2% for DOX and 19.2% for HCPT. Furthermore, the drug release was predominantly at a lower pH owing to the acid labile bond, thereby making the system particularly especially useful for anticancer therapy [7].

One of the key challenges faced while encapsulating drugs in the hydrophobic core of dendrimers is the burst release of drug(s). However, conjugating the drugs is likely to result in very slow release. Thus, for achieving an optimum release, one group of researchers loaded drugs in PLGA nanoparticles and employed a partially cross-linked hydrogel dendrimer as a dispersing agent. Two drugs, i.e., brimonidine and timolol maleate were loaded into the dendrimeric system with the intention of treating glaucoma. The system was found to display high drug loading, a prolonged delivery of drugs (upto one week), and the anti-glaucoma effect lasting for 4 days, all these following a one-time topical application. This highlights the significance of drug loading and release properties [8].

Targeting is an important attribute that must be conferred to dendrimers for the successfully targeting and selective drug delivery to a specific tissue (in other words, a minimal off-target accumulation). Because dendrimers have multiple surface functional groups which can be easily modified to suit the targeting properties, they are favorite candidates for this purpose. Unlike the case with linear polymers, dendrimers may exhibit a high ligand density with step-wise increase in generation numbers, thus facilitating ligand-receptor binding. With the aim of conferring targeting properties, researchers conjugated Apt-ODN (an aptamer-oligo-deoxynucleotide) to the dendrimer. Subsequently, DOX was intercalated with the nucleic acids. This aptamer had been designed for targeting the PSA (prostate specific antigen) which is over-expressed in prostate cancer. The ODN served as an immuno-stimulatory agent. Because the timing and scheduling has to play a critical role during synergism, incorporation of both the immuno-stimulatory as well as the chemotherapeutic agent should result in a stronger therapeutic efficacy. The Apt-ODN complex was found to be stable even after one day, owing to the steric protection provided by the dendrimer. Further, the hybrid (DNA-RNA) demonstrated high drug loading and stimulated both TLR7 and TLR9 [9]. However, this kind of design may be unique to the intercalating properties specifically associated with DOX, thereby indicating that more efforts need to be directed towards the development of a more versatile and robust platform technology, suitable to be applied to any drug combination.

Even though dendrimers offer a useful and exciting platform for co-delivery and combination therapy, these carriers have not crossed the stage of infancy and flourished like other conventional carriers, e.g., liposomes. Plasma half-life and biodegradability are two of the many challenges that still need to be tackled. It is expected that major advances/breakthroughs in dendrimer chemistry particularly with respect to the ease of synthesis, reproducibility and cost-effectiveness should foster further development of dendrimer-based drug delivery including co-delivery [10].

The dendrimers that are used in nanotheranostics are generally 10 to 100 nm [11]. The fifth generation (G5) dendrimers possessing higher hydrophobicity are generally given preference due to enhanced drug stability within the dendrimers [12, 13]. Researchers have reported a replication-deficient adenovirus (serotype 5) bearing the hNIS (sodium-iodide symporter) gene which was coated with PAMAM-G5 and evaluated for its transduction efficacy using ^{123}I-scintigraphy in a xenograft mouse model of hepatocellular carcinoma. Following the dendrimer coating, the replication-deficient serotype 5/hNIS adenovirus could demonstrate partial protection from neutralizing antibodies besides an enhanced transduction efficacy in Coxsackie-adenovirus receptor-negative cells, observed *in vitro*. Further, *in vivo* ^{123}I-scintigraphy of nude mice suggested significantly lower liver transgene expression following intravenous administration of the dendrimer-coated replication-deficient adenovirus [14]. Other workers have designed, characterized, and assessed a theranostic dendrimer both *in vitro* and *in vivo* [15]. These nanocarriers could deliver paclitaxel and the diagnostic agent called Cy5.5. A synthetic analog of LHRH (luteinizing hormone-releasing hormone) targeted to receptors which are overexpressed on the membrane of tumor cells was linked to the dendrimer as a tumor targeting group. Significant differences were noted between the various non-targeted carriers with respect to their tumor distribution, cellular internalization, cytotoxicity, and anticancer efficacy. A higher cellular uptake of the cancer targeted theranostic dendrimer was also accompanied by the targeted accumulation of paclitaxel and this could reduce the adverse effect on healthy tissues besides increasing the specificity as well as sensitivity of tumor imaging by using fluorescent probes. The study clearly demonstrated the theranostic potential of LHRH receptor targeted dendrimer [15].

In another study, the researchers developed dendrimer-based targeted cancer theranostics for the delivery of phthalocyanines (Pc) (Figure 6.1) [16]. The preparation process involved the alteration of Pc with a hydrophobic linker, thus significantly enhancing the entrapment of the hydrophobic drug into a G4 PPIG4 (poly-propylenimine) dendrimer. In order to improve the biocompatibility as well as the tumor-targeted delivery, the surface of Pc-PPIG4 polyplexes was additionally altered with PEG and LHRH, respectively. The Pc derivative encapsulated in the dendrimeric-nanocarrier exhibited a distinct NIR absorption at 700 nm as well as fluorescence emission at 710 and 815 nm, essential prerequisites for an optimum photodynamic therapy and fluorescence imaging. Subsequently, it was shown that the *in vitro* subcellular localization and *in vivo* organ distribution of the so developed nanocarrier could be ascertained from the intrinsic fluorescence properties of the encapsulated Pc. The prepared formulation showed significant photodynamic therapy (up to 1 day) and high cytotoxicity. *In vitro* and *in vivo* imaging studies revealed the efficient internalization capability of the LHRH targeted theranostic dendrimer into cancer cells besides a substantial tumor accumulation. The study thus revealed the remarkable potential of dendrimers as excellent NIR theranostic agents [16].

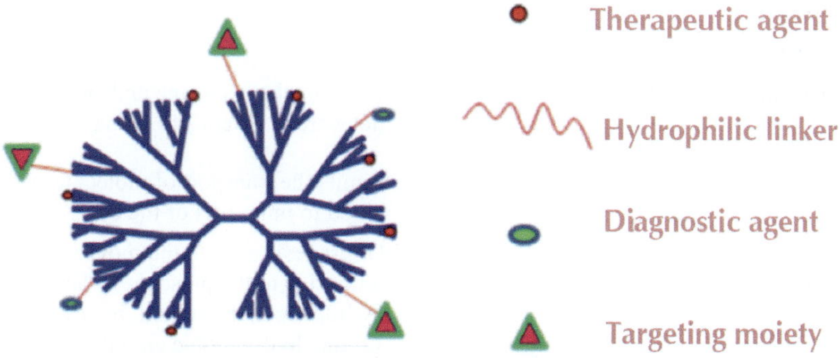

FIGURE 6.1 Concept of a dendrimeric nanotheranostic. (Reprinted with permission fromTaratula, O., Schumann, C., Naleway, M. A., et al. (2013). A multifunctional theranostic platform based on phthalocyanine-loaded dendrimer for image guided drug delivery and photodynamic therapy. Mol Pharm. 10(10), 3946–3958. © 2013. American Chemical Society.)

6.3 OCULAR DRUG DELIVERY

Drug delivery to the eyes remains one of the most challenging tasks in pharmaceutical research. The eye has a complex structure with multilevel barriers (Table 6.1) and thus has high resistance to foreign substances including drugs [17–20]. Both the anterior and posterior segments of the eye respond independently following an ocular application [21]. Thus, ocular/ophthalmic drug delivery may reasonably be categorized into anterior and posterior segments. The conventional drug delivery systems often are not efficient enough to meet the requirements for the treatment of ophthalmic diseases [22]. Even so, >90% of the marketed ophthalmic formulations exist in the form of eye drops, and majority of these actually target the anterior segment eye disorders [23]. Suboptimal bioavailability of drugs from the various ocular dosage forms is largely due to the many pre-corneal 'loss factors' (Table 6.2) which are the main hindrances in anterior segment drug delivery.

The treatment of posterior segment diseases is also an unsolved issue. Majority of the ophthalmic diseases involve the choroid, vitreous and retina. For example, age-related macular degeneration (AMD), glaucoma,

TABLE 6.1 Multilevel Barriers Impeding Ocular Drug Delivery

Barrier	Comments
Inner and outer blood-retinal barriers separate retina and vitreous from the systemic circulation.	Absence of cellular components in vitreous reduces convection of molecules.
Inner limiting membrane.	Controls the exchange and entry of particles from vitreous to retina.
Blood-aqueous barrier.	Limits the transport of molecules from blood to inner part of the eye.
Integrity of corneal epithelium with desmosomes and tight junctions.	Offers resistance to the movement of drugs due to the presence of layers: hydrophobic epithelium, hydrophilic stroma, hydrophobic endothelium.
Tear film	Forms a muco-aqueous barrier continuously washing away the particles at the anterior surface of the eye.

TABLE 6.2 Pre-Corneal 'Loss Factors' Responsible for Suboptimal Bioavailability of Drugs to Anterior Segment from Ocular Dosage Forms

- Solution drainage.
- Lacrymation.
- Tear dynamics.
- Tear dilution.
- Tear turnover.
- Conjunctival absorption.
- Transient residence time in the cul-de-sac.
- Low permeability of the corneal epithelial membrane.

diabetic retinopathy, and many forms of retinitis pigmentosa cause damage the posterior eye segment, leading to impaired vision and even permanent blindness [24]. Drug delivery to the posterior segment is even more challenging compared to the anterior segment, largely due to the acellular nature of the vitreous and the greater diffusion distance [25]. Hence, posterior eye segment disorders have emerged as an important therapeutic target with as yet unmet medical needs. The key aim in the treatment of posterior segments disorders is the optimum delivery of therapeutic drug doses to the tissues vis-a-vis reducing the adverse effects. Various drug delivery systems that have been developed to achieve this aim include liposomes, micelles, nanoparticles, dendrimers, iontophoresis, gene delivery systems, etc. [25–27].

The drug lipophilicity, solubility, size, shape, charge/degree of ionization also influence the ocular penetration rate [28]. The biocompatibility of drug delivery systems is also pertinent as far as ocular delivery is concerned [29, 30]. The criteria for an ideal topical ophthalmic formulation include- (i) the formulation should be well tolerated; (ii) easy to administer; (iii) avoid systemic absorption; and (iv) an enhanced drug retention time in the eye [31, 32]. Presently, nanocarrier-based ophthalmic drug delivery systems including dendrimers offer an appealing way to meet these requirements of a close to ideal ocular therapeutic delivery system.

Dendrimers have been evaluated for ocular drug delivery because they offer several advantages as carrier systems [33]. Table 6.3 gives a list of selected ophthalmic applications of dendrimers.

TABLE 6.3 Selected Ophthalmic Applications of Dendrimers

Dendrimer	Drug	Route of administration	Indication	Reference
PAMAM G1.5–4	Pilocarpine nitrate, tropicamide	Topical	Myosis and mydriasis	34
Phosphorus containing dendrimers	Carteolol	Topical	Glaucoma	35
Dendrimeric polyguanidilyated translocators	Gatifloxacin	Topical	Conjunctivitis and intraocular infections	36
PAMAM G3.5-COOH	Glucosamine and glucosamine 6-sulfate	Subconjunctival	Antiangiogenic in glaucoma surgery	37
PAMAM G3.5-COOH (dendrimeric nanoparticles)	Carboplatin	Subconjunctival	Retinoblastoma	38
Lipophilic amino-acid dendrimer	VEGF-ODN	Intravitreal	CNV	39
PAMAM	Puerarin	Topical	Ocular hypertension and cataract	40
PAMAM G4-OH	Fluocinolone acetonide	Intravitreal	Retinitis	41
Hybrid PAMAM dendrimer hydrogel/PLGA nanoparticle	Brimonidine and timolol maleate	Topical	Gaucoma	42

Various ocular delivery routes have been successfully employed for dendrimer-based drug delivery and improved aqueous solubility, permeability, biocompatibility and bioavailability have been reported. Researchers have evaluated several series of PAMAM dendrimers for controlled drug delivery to the eye. The residence time of tropicamide and pilocarpine nitrate were found to be longer in case of anionic dendrimer solutions. Further, in albino rabbits, the results of miotic activity test on showed that the PAMAM formulations enhanced the pilocarpine nitrate bioavailability as compared to the control and even extended the reduction of IOP (intraocular pressure), thereby suggesting increased precorneal residence time [34].

Investigations with DPTs (dendrimeric poly-guanidilyated translocators, a class of dendrimers having surface guanidine groups and tritolyl branches) as potential carriers for ophthalmic delivery of gatifloxacin revealed that DPTs form stable gatifloxacin complexes and improve the solubility and permeability. In addition, the anti-MRSA activity was also enhanced with a potential delivery system permitting once daily dosing [36].

Phosphorus containing dendrimers (generation 0–2) having a quaternary ammonium compound as the core and a few carboxylic acid terminal groups were synthesized. They were tested *in vivo* in rabbit model for the ocular delivery of carteolol. An increase of carteolol concentration was observed in the aqueous humor. No irritation was noted even after several hours of applying the cationic dendrimers [35].

Researchers synthesized water-soluble conjugates of glucosamine and glucosamine 6-sulfate with G3.5 anionic PAMAM dendrimers aiming to obtain dual properties, i.e., synergistic immunomodulatory and anti-angiogenic effects. When the dendrimeric glucosamine and glucosamine 6-sulfate conjugates were tested together in a clinically pertinent rabbit model of scar tissue formation post-glaucoma filtration surgery, it was observed that the long-term success of surgery had gone up from 30% to 80%. Further, neither microbial infections nor any clinical/hematological/biochemical toxicity were observed in any animal [37].

Intraocular tumors, e.g., retinoblastoma are known to present a high risk of complications including a high metastatic potential. An important study that was performed for drug delivery to intraocular tumors explored the G3.5-COOH PAMAM dendrimers (with carboxyl end groups) for sustained delivery and prolonged half life of carboplatin, and with lowered toxicity. Following subconjunctival administration, these carboplatin-loaded PAMAM dendrimer complexes were evaluated in a transgenic murine retinoblastoma model. Interestingly, the drug-loaded dendrimer nanoparticles not only crossed the sclera but were retained for a prolonged period in the tumor vasculature and provided a sustained therapeutic effect [38].

Biocompatible conjugates of lipophilic amino acid dendrimers with collagen scaffolds have also been developed in order to obtain better physical/mechanical and adhesion properties. The dendrimer-based approach was employed for anti-VEGF-ODN (vascular endothelial growth factor-oligonucleotide) delivery. It was successfully tested in a rat model for treating CNV (choroidal neovascularization). The results suggested that the dendrimer complexes significantly inhibited VEGF expression by 40% to 60%. The injected rat eyes also revealed that no significant toxicity was caused by these complex injections [39].

From the point of view of wound repair, researchers have generated highly cross-linked collagen with G2 poly(propyleneimine)-octaamine dendrimers as tissue-engineering corneal scaffold. The optical transparencies of dendrimer-cross-linked collagen were compared with EDC (1-ethyl-3-(3-dimethyl aminopropyl) carbodiimide hydrochloride) and glutaraldehyde-cross-linked collagen thermal gels. The transparency of dendrimer-cross-linked collagen was noted to be significantly higher. Results further showed that the dendrimer-cross-linked collagen gels boosted human corneal epithelial cell growth and adhesion vis-à-vis avoiding cellular toxicity [43]. The same research group conjugated surface modified dendrimers with cell adhesion peptides as corneal tissue engineering scaffolds. The material was incorporated into the bulk structure of the gels as well as onto the gel surface. It was found that surface modification (dendrimer amine groups modified using carboxyl groups) enhanced human corneal epithelial cell proliferation and adhesion [44]. Other researchers developed a series of dendrimer-based (G1, G2, and G3) adhesives for repairing corneal wounds, conjugated with PEG, glycerol, and succinic acid. Further, the polymer was modified with terminal methacrylate groups ([G1]-PGLSA-MA)$_2$-PEG. Subsequently, two strategies were explored to form the ocular adhesives: (i) a photo-cross-linking reaction; and (ii) a peptide ligation reaction, for coupling together the individual dendrimers. It was observed that both hydrogels were hydrophilic, soft, elastic, transparent, and adhesive. *In vivo* studies in chicken eyes demonstrated that the $([G1]\text{-PGLSA-MA})_2$-PEG adhesive sealed completely on the post-operative day. Further, after 28 days, the histological studies demonstrated the wounds to seal even more completely with these adhesive gels in comparison with sutured wounds. The merit of using photo-cross-linked gels is obviously the ability of the polymer to crosslink and adhere to tissues, following light exposure, even though there is a potential risk of eye damage while using light [45].

Photodynamic therapy is known to be an efficient treatment modality for retinoblastoma besides various other solid tumors. Researchers have designed a photosensitizer, viz. porphyrin-based glyco-dendrimers with surface conjugated Concanavalin A (a mannose-specific ligand protein), for specifically targeting tumor cells in the retina. It was observed that the mannosylated dendrimers exhibited specific interactions with the receptors present in the lipid bilayer. The accumulation in malignant tissue was augmented [33]. Dendrimers have also been evaluated as drug carriers as well as photosensitizers for exudative AMD (age-related macular degeneration) and CNV treatment. Researchers investigated other porphyrin-based dendrimers for their therapeutic efficacy in retinal tumors and exudative AMD associated with CNV. The dendrimers demonstrated selective accumulation in the CNV lesions within 24 h following injection in a CNV rat model [46, 47]. The same workers developed phthalocyanine core-based dendrimeric photosensitizers for compacting and delivering specific therapeutic genes using a targeting approach. The transgene expression was monitored in the irradiated areas only, following subconjunctival injection of the dendrimers, later followed by laser irradiation [48]. Other workers investigated the accumulation of DP (dendrimer porphyrin), DP-encapsulated polymeric micelles, and the efficacy of PDT (photodynamic therapy) using a murine corneal neovascularization model. In this approach, a G3 aryl ether dendrimer zinc porphyrin having carboxyl terminal groups and polymeric micelles made up of DP and PEG-poly(L-lysine) were employed for PDT as a photosensitizer preparation. The results revealed that following administration, both DP and DP-micelle accumulated in the neovascularized area within 1 hour to 24 hours [49].

The *in vitro* and *in vivo* tolerance of G1 and G3 (anionic/cationic) carbosilane dendrimers for topical ophthalmic application were assessed. The formulations when applied to New Zealand albino rabbits revealed that the animals did not present with any discomfort or clinical signs following the administration of the dendrimeric solutions. Nonionic interactions through hydrogen bonding between PAMAM dendrimer surface entities and mucins were noted. The MTT test results further confirmed that anionic dendrimers were nontoxic for conjunctival as well as corneal cells [50].

Recently, puerarin–dendrimer complexes were synthesized using G3.5, G4, G4.5 and G5 PAMAM dendrimers. Their physicochemical characteristics, *in vitro* release, corneal permeation and residence times were determined in rabbits using diffusion cells with excised corneas. It was observed that the puerarin-dendrimers exhibited significantly longer residence time in the rabbit eyes compared with simple puerarin eye drops, without causing damage to the corneal epithelium/endothelium. Further, results of *in vitro* release studies demonstrated puerarin release to be much slower from the complexes in contrast to free puerarin in PBS. However, the corneal permeation studies revealed no significant difference between puerarin-dendrimer complexes versus puerarin eye drops on the drug permeability coefficient [40].

In another study, targeted drug delivery for retinal neuro-inflammation was explored using hydroxyl-terminated G4 PAMAM dendrimer-drug conjugates as nanodevices. Fluocinolone acetonide was linked to the dendrimers and the *in vivo* efficacy was evaluated for 4 weeks, in the Royal College of Surgeons rat retinal degeneration model. It was observed that following intravitreal administration, the PAMAM dendrimers could selectively localize within the activated outer retinal microglia in 2 retinal degeneration rat models. Further, the dendrimers could be detected in the target cells even 35 days post-administration [41].

A PAMAM dendrimer hydrogel was developed by a group of researchers. It comprised of ultraviolet-cured PAMAM dendrimers conjugated with PEG-acrylate chains meant to deliver two antiglaucoma drugs, i.e., 0.1% w/v brimonidine and 0.5% w/v timolol maleate. The dendrimeric hydrogels were obtained by crosslinking the acrylate groups (which are chemically reactive), triggered by UV radiation. It was observed that the dendrimeric hydrogels were mucoadhesive as well as nontoxic to the human corneal epithelial cells. Further, a higher uptake from the epithelial cells and significantly improved corneal transport were noted for both the drugs when compared to the eye drops. This higher uptake with the dendrimeric hydrogel formulations could be explained by the temporary decomposition of the epithelial tight junctions in the cornea [51]. The same group also formulated HDNP, a novel hybrid PAMAM dendrimer hydrogel/PLGA (poly[lactic-co-glycolic acid]) nanoparticle platform, also meant for the co-delivery of brimonidine and timolol maleate. These formulations were again evaluated in terms of the *in vitro* toxicity potential. It was found that the nanoparticles were not cytotoxic to human corneal epithelial cells. Further, in adult normotensive Dutch-belted male rabbits, topical administration of HDNP formulation was effective and could maintain significantly higher concentrations of both the drugs for up to 1 week in aqueous humor as well as cornea when compared to saline. It was also reported that the dendrimeric hydrogels and PLGA nanoparticles did not induce any ocular inflammation/discomfort. This study clearly demonstrated the ability of the new formulation to enhance drug bioavailability and sustained drug activity [42]. Other researchers also developed an *in situ* gel formulation employing Lsy_xCys_y dendrimers for use in cataract incisions (instead of using nylon sutures). They demonstrated that the hydrogel sealant procedure was simple, requiring less surgical time in contrast to conventional suturing and without inflicting additional tissue trauma [52].

The above discussion suggests that despite significant advances, effective treatment of ophthalmic diseases remains a challenge, because of the unique anatomy and physiology of the eye including the presence of ocular barriers. During the last two decades, research in ocular drug delivery has made remarkable progress beginning from the use of conventional formulations (solutions, suspensions, ointments) to the more complex viscosity-enhancing *in situ* gel systems, various inserts, colloidal systems, etc. Given their structural features, majority of the ocular diseases are likely to benefit from dendrimer-based, long-lasting drug delivery systems. It is well known that dendrimers offer practical solutions to key drug delivery issues, e.g., solubility, biodistribution and targeting. Because of the ease to control various features of dendrimers such as size, shape, generation, branching pattern and length as well as surface functionality, these polymers are ideal drug carriers for pharmaceutical applications. Extensive research has shown numerous added benefits of dendrimers as far as ocular drug delivery is concerned (Table 6.4). Thus, even though dendrimeric ocular drug delivery systems are not yet approved for clinical use, the promising preclinical results do encourage researchers to explore the various potential opportunities.

6.4 NOW AND NEXT?

There are already some FDA-approved dendrimer-based products in the market [53–56]:

1. Stratus CS Acute Care (Dade Behring): Contains dendrimer-linked monoclonal antibody, launched for cardiac diagnostic testing.
2. Modified Tomalia-type PAMAM dendrimers, SuperFect (Qiagen): Well-known gene transfection agent available for a diverse range of cell lines.
3. VivaGel: A formulation of polyanionic lysine G4 dendrimers having an anionic surface of naphthalene disulfonate (SPL7013) in a Carbopol gel, showing activity against HIV and HSV, already into Phase III clinical trials by Starpharma, according to FDA requirements. It is also the subject of a license agreement with Durex condoms to be used for condom coating.

Although the number of drugs explored for delivery in dendrimer-based systems is virtually endless, Table 6.5 highlights the selected ones in the past few years. Designing suitable dendrimers and toxicity issues

TABLE 6.4 Added Benefits of Dendrimer-Based Ocular Drug Delivery

• Enhanced corneal residence time of drugs administered topically.
• Target retinal inflammation, providing sustained neuroprotection in retinal degeneration.
• Deliver drugs to the retina following systemic administration.
• Effective as corneal glues, potentially replacing sutures post-corneal surgeries.

TABLE 6.5 Selected Examples of Dendrimer-Based Drug Delivery Systems Explored in Recent Years

Dendrimer	Drug	Features	Indication	Reference
G4-PAMAM	Rifampicin	High stability of rifampicin-PAMAM complex at physiological pH and rapid drug release under acidic medium	Tuberculosis	57
1-(4-carbomethoxypyrrolidone)-terminated PAMAM	Oxacillin	Solvent and generation-dependent complexation	Bacterial infection	58
G2–3 PAMAM-NH_2 and PAMAM-OH	Erythromycin, tobramycin	Enhanced solubility	Bacterial infection	59
G5-PPI	Amphotericin B	pH-sensitive drug release	Leishmaniasis	60
G2.5 mPEG-lysine dendritic micelles	Artemether	Enhanced solubility and stability	Malaria	61
Anionic carbosilane dendrimers with sulfated (G3-S16) and naphthyl sulfonated (G2-NF16) end groups	Zidovudine, efavirenz, tenofovir	Additive/synergistic action of carrier and drug	HIV	62
G0-PAMAM	Zinc phthalocyanine	Aggregation on atheromatous plaques (photodynamic therapy)	Atheresclerotic vascular diseases	63
G5-PAMAM	Simvastatin	Enhanced drug solubility and transepithelial transport	Hypercholesterolemia	64
Silver dendrimer nanocomposites	Silver	Synergistic action of carrier and drug	Inflammation	65
G1- and G2-PAMAM	Dexamethasone	Anti-inflammatory effect	Enhance gene transfection efficiency	66

TABLE 6.5 (Continued)

Dendrimer	Drug	Features	Indication	Reference
Affinity binding peptide functionalized dendrimers	Transforming growth factor-β	Modulate the local delivery and availability of growth factors	Musculoskeletal regeneration	67
PAMAM	Small hairpin RNA	Decreased expression of connective tissue proteins	Myocardial fibrosis	68
PAMAM	Endostatin	Anti-angiogenic combination	Endometriosis	69
G0-G3-PAMAM	Insulin, calcitonin	Rank order of absorption enhancement effect: G3>G2>G1>G0	Pulmonary delivery of hormones	70
Cationic acetylated G5-PAMAM	Epirubicin	Hybrid system for simultaneous imaging and drug delivery	Cancer	71
Hyaluronic acid-grafted PAMAM	Topotecan	Simultaneous long systemic circulation and active tumor targeting	Cancer	72
Amphiphilic dendrimer	Doxorubicin	Supramolecular micelles with large void space for drug accommodation	Cancer	73
mPEG-G3 and mPEG-G4 PAMAM	Cisplatin	Binding in hydrophilic mode	Cancer	74

are some of the major hurdles which prevent dendrimer-based drug delivery systems from being translated from basic research to clinically useful products (Chapter 10). It is expected that as research progresses and theses issues are better tackled, more and more such systems would begin to witness the light of the day.

KEYWORDS

- **anti-inflammatory drugs**
- **doxorubicin**
- **host-guest**
- **ocular drugs**
- **prostate specific antigen**
- **supramolecular chemistry**
- **transdermal delivery**

REFERENCES

1. Maciejewski, M. (1982). Concepts of trapping topologically by shell molecules. *J Macromol Sci Chem A 17,* 689.
2. Herrmann, A., Mihov, G., Vandermeulen, G. W. M., Klok, H.-A, & Mullen, K. (2003). Peptide-functionalized polyphenylene dendrimers. *Tetrahedron 59*, 3925.
3. Cheng, Y., Man, N., Xu, T., Fu, R., Wang, X., Wang, X., & Wen, L. (2007). Transdermal delivery of nonsteroidal anti-inflammatory drugs mediated by polyamidoamine (PAMAM) dendrimers. *J Pharm Sci 96*, 595–602.
4. Megan E. Godsey, Smruthi Suryaprakash, & Kam W. Leong. (2013). Materials innovation for co-delivery of diverse therapeutic cargos. *RSC Adv. 3*(47), 24794–24811.
5. Clementi, C., Miller, K., Mero, A., Satchi-Fainaro, R., & Pasut, G. (2011). *Mol. Pharm. 8,* 1063–1072.
6. Pandey, R., & Ahmad, Z. (2011). Nanomedicine and experimental tuberculosis: Facts, flaws and future. *Nanomedicine: Nanotechnology, Biology and Medicine 7*(3), 259–272.
7. Zhang, Y., Xiao, C., Li, M., Chen, J., Ding, J., He, C., Zhuang, X., & Chen, X. (2013). *Macromol. Biosci. 13*(5), 584–594.
8. Yang, H., Tyagi, P., Kadam, R. S., Holden, C. A., & Kompella, U. B. (2012). *ACS Nano. 6*, 7595–7606.

9. Lee, I.-H., An, S., Yu, M. K., Kwon H-K, Im, S-H, & Jon, S. J. (2011). *Controlled Release. 155*, 435–441.
10. Cheng, Y., Zhao, L., Li, Y., & Xu, T. (2011). *Chem. Soc. Rev. 40*, 2673–2703.
11. Fahmy, T. M., Fong, P. M., Park, J., et al. (2007). Nanosystems for simultaneous imaging and drug delivery to T Cells. *AAPS J. 9*(2), E171–E180.
12. Frechet, J. M. (1994). Functional polymers and dendrimers – reactivity, molecular architecture, and interfacial energy. *Science. 263*(5154), 1710–1715.
13. Li, Y., Cheng, Y., & Xu, T. (2007). Design, synthesis and potent pharmaceutical applications of glycodendrimers; a mini review. *Curr Drug Discov Technol. 4*(4), 246–254.
14. Grünwald, G. K., Vetter, A., Klutz, K. et al. (2013). Systemic image-guided liver cancer radiovirotherapy using dendrimer-coated adenovirus encoding the sodium iodide symporter as theranostic gene. *J Nucl Med. 54*(8), 1450–1457.
15. Saad, M., Garbuzenko, O. B., Ber, E., et al. (2008). Receptor targeted polymers, dendrimers, liposomes: which nanocarrier is the most efficient for tumor-specific treatment and imaging? *J Control Release. 130*(2), 107–114.
16. Taratula, O., Schumann, C., Naleway, M. A., et al. (2013). A multifunctional theranostic platform based on phthalocyanine-loaded dendrimer for image guided drug delivery and photodynamic therapy. *Mol Pharm. 10*(10), 3946–3958.
17. Diebold, Y., & Calonge, M. (2010). "Applications of nanoparticles in ophthalmology," *Progress in Retinal and Eye Research, 29*(6), 596–609.
18. Yasukawa, T., Ogura, Y., Tabata, Y., Kimura, H., Wiedemann, P., & Honda, Y. (2004). "Drug delivery systems for vitreoretinal diseases," *Progress in Retinal and Eye Research, 23*(3), 253–281.
19. Campbell, M., Nguyen, A. T. H., Kiang, A. S. et al. (2010). "Reversible and size-selective opening of the inner blood-retina barrier: a novel therapeutic strategy," in *Retinal Degenerative Diseases: Laboratory and Therapeutic Investigations, vol. 664*, 301–308, Springer, New York, NY, USA.
20. Pahuja, P., Arora, S., & Pawar, P. (2012). "Ocular drug delivery system: a reference to natural polymers," *Expert Opinion on Drug Delivery, 9*(7), 837–861.
21. Nagarwal, R. C., Kant, S., Singh, P. N., Maiti, P., & Pandit, J. K. (2009). "Polymeric nanoparticulate system: a potential approach for ocular drug delivery," *Journal of Controlled Release, 136*(1), 2–13.
22. Gaudana, R., Jwala, J., Boddu, S. H. S., & Mitra, A. K. (2009). "Recent perspectives in ocular drug delivery," *Pharmaceutical Research, 26*(5), 1197–1216.
23. Lang, J. C. (1995). "Ocular drug delivery conventional ocular formulations," *Advanced Drug Delivery Reviews, 16*(1), 39–43.
24. Ranta, V. P., Mannermaa, E., Lummepuro, K. et al. (2010). "Barrier analysis of periocular drug delivery to the posterior segment," *Journal of Controlled Release, 148*(1), 42–48.
25. Eljarrat-Binstock, E., Peer, J. & Domb, A. J. (2010). "New techniques for drug delivery to the posterior eye segment," *Pharmaceutical Research, 27*(4), 530–543.
26. Del Amo, E. M. & Urtti, A. (2008). "Current and future ophthalmic drug delivery systems: a shift to the posterior segment," *Drug Discovery Today, 13*(3–4), 135–143.
27. Ranta, V. P., & Urtti, A. (2006). "Transscleral drug delivery to the posterior eye: prospects of pharmacokinetic modeling," *Advanced Drug Delivery Reviews, 58*(11), 1164–1181.

28. Wadhwa, S., Paliwal, R., Paliwal, S. R., & Vyas, S. P. (2009). "Nanocarriers in ocular drug delivery: an update review," *Current Pharmaceutical Design, 15*(23), 2724–2750.
29. Raghava, S., Hammond, M., & Kompella, U. B. (2004). "Periocular routes for retinal drug delivery," *Expert Opinion on Drug Delivery, 1*(1), 99–114.
30. Patel, P. B., Shastri, D. H., Shelat, P. K., & Shukla, A. K. (2010). "Ophthalmic drug delivery system: challenges and approaches," *Systematic Reviews in Pharmacy, 1*(2), 113–120.
31. Perry, H. D., Solomon, R., E. D. Donnenfeld et al. (2008). "Evaluation of topical cyclosporine for the treatment of dry eye disease," *Archives of Ophthalmology, 126*(8), 1046–1050.
32. Yavuz, B., Bozdag, S., Pehlıvan, & Unlu, N. (2012). "An overview on dry eye treatment: approaches for cyclosporin a delivery," *The Scientific World Journal, vol. 2012*, 1–12.
33. Burçin Yavuz, Sibel Bozdag Pehlivan, Nursen Ünlü. Dendrimeric Systems and Their Applications in Ocular Drug Delivery. The Scientific World Journal Volume 2013, Article ID 732340.
34. Vandamme, T. F., & Brobeck, L. (2005). "Poly (amidoamine) dendrimers as ophthalmic vehicles for ocular delivery of pilocarpine nitrate and tropicamide," *Journal of Controlled Release, 102*(1), 23–38.
35. Spataro, G. G., Malecaze, F., Turrin, C.-O. et al. (2010)."Designing dendrimers for ocular drug delivery," *European Journal of Medicinal Chemistry, 45*(1), 326–334.
36. Durairaj, C., Kadam, R. S., Chandler, J. W., Hutcherson, S. L., & Kompella, U. B. (2010). "Nanosized dendritic polyguanidilyated trans-locators for enhanced solubility, permeability and delivery of gatifloxacin," *Investigative Ophthalmology and Visual Science, 51*(11), 5804–5816.
37. Shaunak, S., Thomas, S., E. Gianasi et al. (2004). "Polyvalent dendrimer glucosamine conjugates prevent scar tissue formation," *Nature Biotechnology 22*(8), 977–984.
38. Kang, S. J., Durairaj, C., Kompella, U. B., O'Brien, J. M. & Grossniklaus, H. E. (2009). "Subconjunctival nanoparticle carboplatin in the treatment of murine retinoblastoma," *Archives of Ophthalmology, 127*(8), 1043–1047.
39. Marano, R. J., Toth, I., Wimmer, N., Brankov, M., & Rakoczy, P. E. (2005). "Dendrimer delivery of an anti-veg oligonucleotide into the eye: a long-term study into inhibition of laser-induced CNV, distribution, uptake and toxicity," *Gene Therapy, 12*(21), 1544–1550.
40. Yao, W. J., Sun, K. X., H. J. Mu et al. (2010). "Preparation and characterization of puerarin dendrimer complexes as an ocular drug deli-very system," *Drug Development and Industrial Pharmacy, 36*(9), 1027–1035.
41. Iezzi, R., Guru, B. R., Glybina, I. V., Mishra, M. K., Kennedy, A., & Kannan, R. M. (2012). "Dendrimer-based targeted intravitreal therapy for sustained attenuation of neuroinflammation in retinal degeneration," *Biomaterials, 33*(3), 979–988.
42. Yang, H., Tyagi, P., Kadam, R. S., Holden, C. A., & Kompella, U. B. (2012). "Hybrid dendrimer hydrogel/plga nanoparticle platform sustains drug delivery for one week and antiglaucoma effects for four days following one-time topical administration," *ACS Nano, 6*(9), 7595–7606.

43. Duan, X., & Sheardown, H. (2006). "Dendrimer cross-linked collagen as a corneal tissue engineering scaffold: mechanical properties and corneal epithelial cell interactions," *Biomaterials, 27*(26), 4608–4617.
44. Duan, X. D., McLaughlin, C., Griffith, M., & Sheardown, H. (2007). "Biofunctionalization of collagen for improved biological response: scaffolds for corneal tissue engineering," *Biomaterials, 28*(1), 78–88.
45. Grinstaff, M. W. (2007). "Designing hydrogel adhesives for corneal wound repair," Biomaterials, 28(35), 5205–5214.
46. Nishiyama, N., Stapert, H. R., G. D. Zhang et al. (2003). "Light-harvesting ionic dendrimer porphyrins as new photosensitizers for photodynamic therapy," *Bioconjugate Chemistry, 14*(1), 58–66.
47. Nishiyama, N., Morimoto, Y., Jang, W. D., & Kataoka, K. (2009). "Design and development of dendrimer photosensitizer-incorporated polymeric micelles for enhanced photodynamic therapy," *Advanced Drug Delivery Reviews, 61*(4), 327–338.
48. Nishiyama, N., Iriyama, A., Jang, W. D. et al. (2005). "Light-induced gene transfer from packaged DNA enveloped in a dendrimeric photosensitizer," *Nature Materials, 4*(12), 934–941.
49. Sugisaki, K., Usui, T., Nishiyama, N. et al. (2008). "Photodynamic therapy for corneal neovascularization using polymeric micelles encapsulating dendrimer porphyrins," *Investigative Ophthalmology and Visual Science, 49*(3), 894–899.
50. Bravo-Osuna, I., Herrero-Vanrell, R., Molina Martınez, I. T. et al. (2010). "In vitro and in vivo tolerance studies of carbosilane dendrimers for ophthalmic administration," *Investigative Ophthalmologyand Visual Science, 51*, E-Abstract 437.
51. Holden, C. A., Tyagi, P., Thakur, A. et al. (2012). "Polyamidoamine dendrimer hydrogel for enhanced delivery of antiglaucoma drugs," *Nanomedicine, 8*(5), 776–783.
52. Wathier, M., Jung, P. J., Carnahan, M. A., Kim, T., & Grinstaff, M. W. (2004). "Dendritic macromers as in situ polymerizing biomate-rials for securing cataract incisions," *Journal of the American Chemical Society, 126*(40), 12744–12745.
53. Kumar, P., Meena, K. P., Kumar, P., Choudhary, C., Thakur, D. S., & Bajpayee, P. (2010). "Dendrimer: a novel polymer for drug delivery," *JITPS Journal, 1*(6), 252–269.
54. Menjoge, A. R., Kannan, R. M., & Tomalia, D. A. (2010). "Dendrimer-based drug and imaging conjugates: design considerations for nanomedical applications," *Drug Discovery Today, 15*(5–6), 171–185.
55. McCarthy, T. D., Karellas, P., S. A. Henderson et al. (2005). "Dendrimers as drugs: discovery and preclinical and clinical development of dendrimer-based microbicides for HIV and STI prevention," *Molecular Pharmaceutics, 2*(4), 312–318.
56. Wijagkanalan, W., Kawakami, S., & Hashida, M. (2011). "Designing dendrimers for drug delivery and imaging: pharmacokinetic considerations," *Pharmaceutical Research, 28*(7), 1500–1519.
57. Bellini, R. G., Guimarães, A. P., Pacheco, M. A., Dias, D. M., Furtado, V. R., de Alencastro, R. B., & Horta, B. A. (2015). Association of the anti-tuberculosis drug rifampicin with a PAMAM dendrimer. *J Mol Graph Model. 60*, 34–42.
58. Hansen, J. S., Ficker, M., Petersen, J. F., Nielsen, B. E., Gohar, S., & Christensen, J. B. (2013). Study of the complexation of oxacillin in 1-(4-carbomethoxypyrrolidone)-terminated PAMAM dendrimers. *J Phys Chem, B. 117*(47), 14865–14874.

59. Winnicka, K., Wroblewska, M., Wieczorek, P., Sacha, P. T., & Tryniszewska, E. A. (2013). The effect of PAMAM dendrimers on the antibacterial activity of antibiotics with different water solubility. *Molecules. 18*(7), 8607–8617.
60. Jain, K., Verma, A. K., Mishra, P. R., & Jain, N. K. (2015). Surface-engineered dendrimeric nanoconjugates for macrophage-targeted delivery of amphotericin, B. formulation development and in vitro and in vivo evaluation. *Antimicrob Agents Chemother. 59*(5), 2479–2487.
61. Bhadra, D., Bhadra, S., & Jain, N. K. (2005). Pegylated lysine based copolymeric dendritic micelles for solubilization and delivery of artemether. *J Pharm Pharm Sci 8*(3), 467–482.
62. Vacas-Córdoba, E., Galán, M., de la Mata, F. J., Gómez, R., Pion, M., Muñoz-Fernández, M. Á. (2014). Enhanced activity of carbosilane dendrimers against HIV when combined with reverse transcriptase inhibitor drugs: searching for more potent microbicides. *Int J Nanomedicine. 9*, 3591–3600.
63. Spyropoulos-Antonakakis, N., Sarantopoulou, E., Trohopoulos, P. N., Stefi, A. L., Kollia, Z., Gavriil, V. E., Bourkoula, A., Petrou, P. S., Kakabakos, S., Semashko, V. V., Nizamutdinov, A. S., & Cefalas, A. C. (2015). Selective aggregation of PAMAM dendrimernanocarriers and PAMAM/ZnPc nanodrugs on human atheromatous carotid tissues: a photodynamic therapy for atherosclerosis. *Nanoscale Res Lett. 10*, 210.
64. Qi, R., Zhang, H., Xu, L., Shen, W., Chen, C., Wang, C., Cao, Y., Wang, Y., van Dongen, M. A., He, B., Wang, S., Liu, G., BanaszakHoll, M. M., & Zhang, Q. (2015). G5 PAMAM dendrimer versus liposome: A comparison study on the in vitro transepithelial transport and in vivo oral absorption of simvastatin. *Nanomedicine. 11*(5), 1141–1151.
65. Liu, X., Hao, W., Lok, C. N., Wang, Y. C., Zhang, R., & Wong, K. K. (2014). Dendrimer encapsulation enhances anti-inflammatory efficacy of silver nanoparticles. *J Pediatr Surg. 49*(12), 1846–1851.
66. Kim, J. Y., Ryu, J. H., Hyun, H., Kim, H. A., Choi, J. S., Yun Lee, D., Rhim, T., Park, J. H., & Lee, M. (2012). Dexamethasone conjugation to polyamidoamine dendrimers G1 and G2 for enhanced transfection efficiency with an anti-inflammatory effect. *J Drug Target. 20*(8), 667 677.
67. Seelbach, R. J., Fransen, P., Pulido, D., D'Este, M., Duttenhoefer, F., Sauerbier, S., Freiman, T. M., Niemeyer, P., Albericio, F., Alini, M., Royo, M., Mata, A., & Eglin, D. (2015). Injectable Hyaluronan Hydrogels with Peptide-Binding Dendrimers Modulate the Controlled Release of BMP-2 and TGF-β1. *Macromol Biosci.* May 5. doi: 10.1002/mabi.201500082. [Epub ahead of print] PMID: 25943094.
68. Huang, Z. J., Yi, B., Yuan, H., Yang, G. P. (2014). Efficient delivery of connective tissue growth factor shRNA using PAMAM nanoparticles. *Genet Mol Res. 13*(3), 6716–6723.
69. Wang, N., Liu, B., Liang, L., Wu, Y., Xie, H., Huang, J., Guo, X., Tan, J., Zhan, X., Liu, Y., Wang, L., & Ke, P. (2014). Antiangiogenesis therapy of endometriosis using PAMAM as a gene vector in a noninvasive animal model. *Biomed Res Int. 2014*, 546479.
70. Dong, Z., Hamid, K. A., Gao, Y., Lin, Y., Katsumi, H., Sakane, T., & Yamamoto, A. (2011). Polyamidoamine dendrimers can improve the pulmonary absorption of insulin and calcitonin in rats. *J Pharm Sci. 100*(5), 1866–1878.

71. Matai, I., Sachdev, A., & Gopinath, P. (2015). Self-assembled hybrids of fluorescent carbon dots and PAMAM dendrimers for epirubicin delivery and intracellular imaging. *ACS Appl Mater Interfaces.* *7*(21), 11423–11435.
72. Qi, X., Fan, Y., He, H., & Wu, Z. (2015). Hyaluronic acid-grafted polyamidoamine dendrimers enable long circulation and active tumor targeting simultaneously. *Carbohydr Polym.* *126*, 231–239.
73. Wei, T., Chen, C., Liu, J., Liu, C., Posocco, P., Liu, X., Cheng, Q., Huo, S., Liang, Z., Fermeglia, M., Pricl, S., Liang, X. J., Rocchi, P., & Peng, L. (2015). Anticancer drug nanomicelles formed by self-assembling amphiphilic dendrimer to combat cancer drug resistance. *Proc Natl Acad Sci USA.* *112*(10), 2978–2983.
74. Abderrezak, A., Bourassa, P., Mandeville, J. S., Sedaghat-Herati, R., & Tajmir-Riahi, H. A. (2012). Dendrimers bind antioxidant polyphenols and cisplatin drug. *PLoS One.* *7*(3), e33102.

CHAPTER 7

DENDRIMERS AND GENE DELIVERY

ZAHOOR AHMAD PARRY, PhD, and RAJESH PANDEY, MD

CONTENTS

7.1 INTRODUCTION

Gene therapy is a proven tool to tackle many suitable selected stubborn diseases such as cancer and genetic/hereditary disorders in which conventional treatments may be ineffective. In addition to DNA transfer, RNA interference (RNAi) has emerged as another approach for gene therapy or gene manipulation. Since its initial report in 1998, RNAi quickly became a powerful tool in basic research as well as to develop novel therapeutics [1–3]. In the general RNAi process, a dsRNA is introduced into a target cell. One strand of the dsRNA is designed to be an antisense RNA because its sequence is complementary to the RNA transcript of the gene selected for silencing. Inside the cell, an ATP-dependent RNAase III called 'dicer,' catalyzes the cleavage of both strands to produce a double stranded small interfering RNA (ds-siRNA), 21–23 nucleotides long and having 2-nucleotides long 3'-overhangs on each strand. The ds-siRNA is passed on to another protein complex called RNA-induced silencing complex (RISC). In an ATP-dependent manner, RISC unwinds the ds-siRNA and selects the antisense strand, now called the 'guide strand.' The other strand called the 'passenger strand' is discarded. RISC pairs the single guide strand with a complementary region on the target gene transcript (mRNA). The RNase H activity of RISC performs its 'slicer' function by cleaving the RNA transcript so that it can no longer be translated by ribosomes. The guide strand remains associated with RISC, hence it can be used for multiple cycles of mRNA cleavage, i.e., post-transcriptional gene silencing [4]. This process is also called post-transcriptional gene silencing and is highly efficient and specific. This is because a single nucleotide mismatch between the target mRNA and the siRNA would prevent the recognition/pairing and thereby the silencing process.

For the purpose of clinical applications, the successful development of RNAi depends on the use of efficient and safe vectors so as to deliver siRNAs into the target cells, by overcoming the extracellular and intracellular barriers. There are some key requirements to achieve this (Table 7.1) [5].

Because of the fact that DNA and siRNA share many structural and physicochemical properties, logically, the vectors developed for DNA could also be applied for delivering siRNA. Like in case of DNA-based gene transfer, the virus-derived carriers have demonstrated high efficiency for delivering siRNA to the host cells by exploiting the intracellular trafficking machinery. However, due to several constraints such as the high cost of production and safety issues (as in DNA transfer), the non-viral

TABLE 7.1 Key Requirements for siRNA Delivery

- Protecting siRNA from degradation in the systemic circulation.
- Transporting siRNA to target sites, avoiding nonspecific delivery
- Promoting cellular uptake.
- Ensuring endosomal escape.
- Releasing the siRNA.
- Making the siRNA readily accessible to the cellular RNAi machinery.

siRNA vectors have attracted increasing attention. These vectors generally have a cationic nature and the examples include the cell penetrating peptides, lipids, polymers, dendrimers. Each of these complex the siRNA via electrostatic interactions [6]. Out of these cationic vectors, the dendrimers are a special class of molecules because they are synthesized/assembled step by step and as a result their architecture may be designed and controlled with precision. As in case of drug delivery, the typical structure of a dendrimer is advantageous for gene delivery because it contains three parts: (i) a central core defining the interior size and the number/direction of the branches; (ii) the repetitive branch units regulating the generation (G), overall size and flexibility; and (iii) the terminal groups defining the chemical properties and the possibility of interactions. Till date, plenty of research work involving the non-viral siRNA delivery has used dendrimers as vectors owing to their unique and flexible chemistry. The many types of dendrimers that have been evaluated for siRNA delivery are shown in Table 7.2.

In 2007, researchers reported the designing/preparation of iNOPs (interfering nanoparticles) as siRNA delivery reagents. The iNOPs had two characteristic subunits: (i) a functionalized dendrimer-lipid nanoparticle as carrier (Figure 7.1a); and (ii) an siRNA (chemically modified) for silencing. Following the injection of iNOPs to mice, gene silencing could be observed for the gene for apolipoprotein B. Further, this iNOP treatment did not elicit an immune response [31]. In 2008, another group of researchers described the preparation of bis-guanidinium-tetrakis-cyclodextrin tetrapod (Figure 7.1b) and the capability of this compound to bind

TABLE 7.2 Dendrimers Evaluated for siRNA Delivery

Types of dendrimers	References
Poly(amidoamine)	7–11
Poly(propylene imine)	12–15
Carbosilane	16–18
Poly(L-lysine)	19–21
Triazine	22–24
Polyglycerol-based	25–27
Nanocarbon-based	28–30
Miscellaneous	25, 31–34

to siRNA. Fluorescence microscopy confirmed the cellular transfection of siRNA into the human embryonic lung fibroblasts [32]. Later, other workers developed dendrimer-conjugated magneto-fluorescent nanoworms (these were designated as dendriworms) as a basis for siRNA delivery. The cross-linked, amine-modified, iron oxide nanoworms were linked with sulfhydryl-containing PAMAM dendrons by using the hetero-bifunctional linker, viz. N-succinimidyl 3-(2-pyridyldithio)-propionate. This dendriworm complex bearing siRNA against the EGFR (epidermal growth factor receptor) couldbetter reduce the protein levels (i.e., EGFR) in human glioblastoma cells, compared with the commercial cationic lipids [33]. Other researchers have reported the preparation of PEI-derived branched structures with PEI-PAMAM (polyamidoamine) or PEI-Glu (gluconolactone) as the shell (Figures 7.1c and 7.1d). It was observed that PEI-Glu was more biocompatible than PEI-PAMAM. Further, the siRNA-PEI-Glu complexes could successfully accumulate in the cytosol in a time-dependent fashion. However, the efficiency of gene silencing was higher with PEI-PAMAM than PEI-Glu [25]. Recently, phosphorus-containing dendrimer for siRNA delivery have also been evaluated. By employing specific siRNA directed against HIV-1 Nef, the dendriplex (siRNA/dendrimer) demonstrated efficient Nef silencing, thus interfering with HIV-1 replication [34].

7.2 ADVANTAGES OF siRNA FOR THERAPY

In general, RNAi is considered an effective strategy for disease management as compared to conventional therapeutics owing to its potential for repressing the translation of any disease-causing protein through gene silencing. Watson-Crick base pairing of RNA with the target sequence can specifically discriminate the target from non-targets [35]. Out of the various RNA molecules (such as siRNA, shRNA and miRNA) that may cause gene-specific silencing through the RNAi pathway, antisense strand of siRNA is fully complementary to the mRNA target. Thus, it has better target recognition and binding compared with other RNAs which are only partially complementary to the target mRNA [35, 36]. The present improved understanding of the subtle molecular pathogenesis of many diseases as well as the regulation of disease-specific biomarkers has made

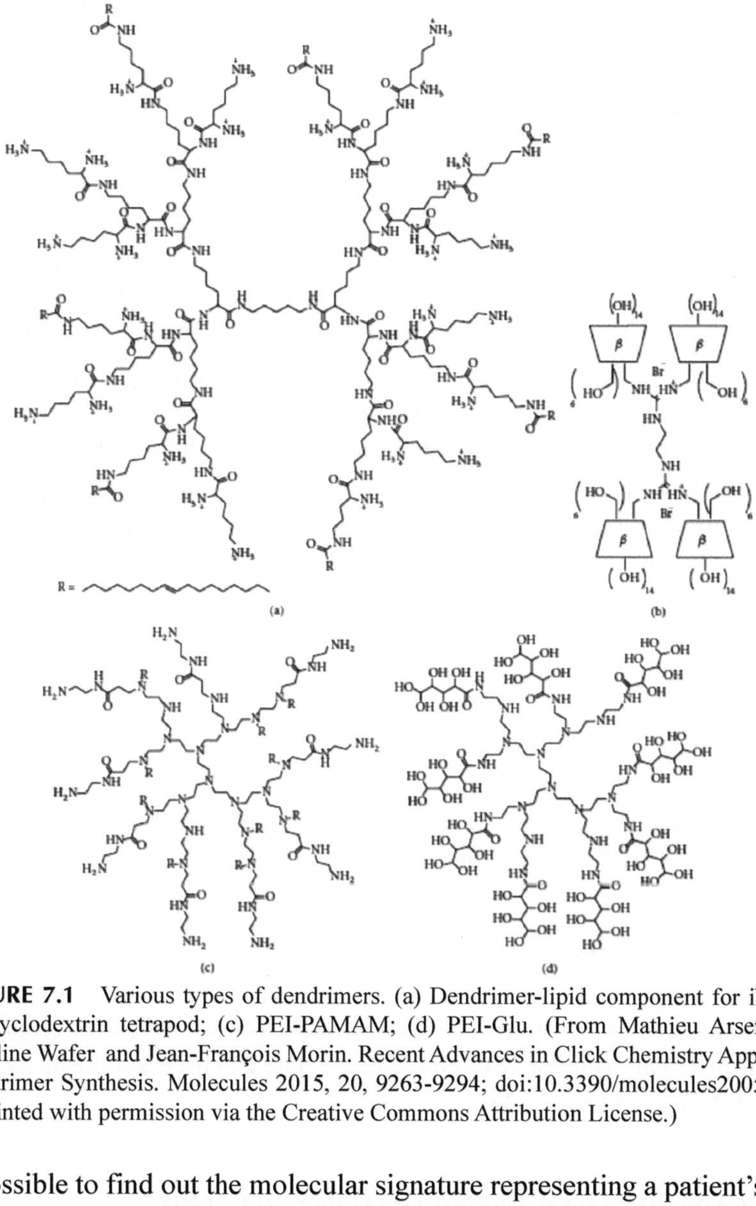

FIGURE 7.1 Various types of dendrimers. (a) Dendrimer-lipid component for iNOPs; (b) cyclodextrin tetrapod; (c) PEI-PAMAM; (d) PEI-Glu. (From Mathieu Arseneault, Caroline Wafer and Jean-François Morin. Recent Advances in Click Chemistry Applied to Dendrimer Synthesis. Molecules 2015, 20, 9263-9294; doi:10.3390/molecules20059263. Reprinted with permission via the Creative Commons Attribution License.)

it possible to find out the molecular signature representing a patient's particular disease. This would enable the development of a customized therapy which would not only be more efficacious but also have a better safety profile [37]. A complete understanding of the siRNA synthesizing methodology enables the preparation of the desired siRNA that would selectively target and turn-off a disease-causing gene. Theoretically, antisense therapeutics may be designed based on the target mRNA sequence and, therefore, any gene could be silenced using this antisense technology [5]. It is not surprising that due to the high potency and least off-targeting of siRNA out of all the candidate antisense molecules, the siRNA has surfaced as the most promising tool to be applied for personalized medicine [38].

Gene silencing based on RNAi technology has distinct advantages over the conventional therapeutics. The inhibitory action of many conventional drugs is mediated largely by blocking a protein's function via binding to its active site. This is the site where the receptor-ligand interaction occurs and leads to downstream activation/inhibition of signaling pathways. However, small molecule-mediated therapeutic action may not be achieved in case the disease-related protein has an unfavorable conformation (that is inaccessible to small molecules), i.e., what is termed a 'non-druggable' target [37–39]. The design as well as synthesis of small molecule ligands for inhibiting a target protein is rather challenging. There is always a possibility of discovering lead molecules, novel ligands and optimization for achieving a higher degree of inhibition compared to an existing ligand. On the contrary, RNAi technology allows the blocking of gene expression (concerning the target protein) rather than actually blocking the activity of the protein of interest [40, 41]. Synthesizing RNA molecules complementary to a particular gene is relatively easy and may be applied by choosing well-established strategies in contrast to the much variable small molecule synthesis.

7.3 CHALLENGES

The therapeutic applications of RNA molecules (i.e., siRNA for all practical purposes) are riddled with challenges related to their synthesis as well as their delivery (Table 7.3).

7.4 DENDRIMERS AS siRNA DELIVERY VECTORS

Dendrimers are one of the most widely studied cationic polymers for nucleic acid delivery. Nevertheless, as mentioned above, the issues regarding delivery of siRNA in particular are relatively unexplored.

TABLE 7.3 Challenges in the Therapeutic Applications of siRNA

Challenges	Potential remedies	Comments
Naked siRNA is highly unstable; following systemic administration, it is rapidly degraded by serum nucleases [35, 42].	• Chemical modification in the siRNA [50]. The 2'-OH group of ribose in the RNA is not essential for siRNA recognition by RNAi machinery, hence extensive modifications at 2'-position of both the strands can be done, and are called 2'-modifications. Examples include an ether linkage as a 2'-methylnucleoside (2'-O-Me) and replacement of the -OH with its bioisostere, F, as in 2-deoxyfluridine (2'-F) [43]. • Another 2'-modification involves introduction of a methylene group between 2'- and 4'-positions by an ether linkage, termed a locked nucleic acid (LNA) modification [44]. • The modifications on the phosphate backbone at 3'-end provide another effective approach, e.g., boranophosphate, phosphothionate, phosphoroamidate and methylphosphonate modifications may be made between two riboses [42, 45–47].	• The modification should be tolerated by the RNAi machinery. • Chemical modifications of siRNA by linking bulky groups at 2'-position may adversely alter the interaction with Dicer and loading into RISC [42, 48].
Naked, unmodified RNA (especially <50 KDa [~10 nm]) undergoes rapid renal clearance. [49, 50].	Nanoparticle-mediated siRNA delivery [51]. Complexing anionic siRNA with cationic polymers or nanoparticles forms the basis for vector-mediated siRNA delivery.	Potential role for dendrimer-based siRNA delivery
Naked siRNA may induce immune response which may lead to the production of pro-inflammatory cytokines [52, 53].	Nanoparticle-mediated siRNA delivery [51].	Potential role for dendrimer-based siRNA delivery

TABLE 7.3 (Continued)

Challenges	Potential remedies	Comments
siRNA may exhibit off-target gene silencing when other mRNA transcripts partially hybridize with the administered siRNA [35, 54]. Various siRNA sequences against the same gene could generate similar gene silencing signatures.	Nanoparticle-mediated siRNA delivery [51].	• Potential role for dendrimer-based siRNA delivery. However, an important difference between the vector-mediated delivery of DNA and siRNA is that due to the smaller size of siRNA, it interacts less efficiently with cationic polymers [55]. • siRNA is less flexible compared to plasmid DNA [56]. Unlike plasmid DNA, a 21–23 base-paired rigid rod-like siRNA does not compact well when complexed with a vector and may result in incomplete encapsulation forming undesirably large complexes [57–59].

Several formulations of dendrimer-DNA complexes have been investigated such as polyethylene glycol-modified PAMAM-DNA, PAMAM-PEG-PAMAM-DNA, PAMAM-DNA, and PPI-DNA, PEI-DNA, etc. [60–64]. The relatively rigid siRNA faces a less efficient interaction with cationic polymers (when compared to DNA). This suggests that a cationic polymer otherwise efficient for DNA delivery may not necessarily be an equally effective siRNA carrier.

To clarify the molecular mechanism of self-assembly of PAMAM-siRNA dendriplex, researchers studied various generations of PMAM dendrimers [65]. G4 and G7 were found to display an equal efficiency as far as dendriplex formation was concerned. However, G1 bearing a lower charge density lacked the ability of siRNA condensation. Smaller mean size and higher polydispersity at an increased dendrimer concentration suggested a thermodynamically favorable electrostatic interaction. A spontaneous, exothermic complexation for G1, and a biphasic, initially exothermic but secondary endothermic complexation for G4 and G7 were observed during the dendrimer-siRNA aggregation process. The flexible G1 and rigid G7 exhibited an entropic penalty, which made G4 the most suitable for dendriplex formation, having a charge density favorable for siRNA binding.

7.5 DENDRIMERS WITH VARIED STRUCTURES FOR siRNA DELIVERY

Dendrimers with variable cores and branching structures have been explored for siRNA delivery [55, 66–69]. Polymerized and PEG-based dendrimers with core-shell structures including polyglyceryl pentaethylenehexamine carbamate, polyglycerolamine (PG-Amine), PEI-gluconolactone and PEI-PAMAM were synthesized and tested for their efficacy as siRNA carriers. A study with PEI-PAMAM dendrimers indicated their low toxicity, besides strongly enhancing the stability of the siRNA and its intracellular trafficking. Further, it demonstrated both *in vitro* and *in vivo* gene silencing. To be more specific, the dendritic polymers showed high efficacy in luciferase gene silencing, the gene being over-expressed in murine mammary adenocarcinoma and human glioblastoma cells. Further, PG-amine exhibited the best silencing efficacy/toxicity ratio. Following the intratumoral and intravenous administration of siRNA-PG-amine polyplexes targeting the luciferase in tumor-bearing mice led to a significant gene silencing within 24 h. *In vivo*, a high degree of luciferase gene silencing could be accomplished within 24 h of treatment with luciferase siRNA in U87-Luc human glioblastoma cells implanted subcutaneously in DA3-mCherry-Luc cells in BALB/C mice and SKID mice.

A family of triazine dendrimers with varying core flexibility, generation as well as surface functionality was assessed for its potential to condense and efficiently deliver a luciferase targeting siRNA aimed at reporter gene knockdown [55]. The triazine groups introduced in the periphery were linked via specific diamine groups. It was observed that the triazine dendrimers, which were otherwise the most efficient DNA delivery vectors, failed to mediate gene silencing. On the other hand, the modestly effective gene delivery vector could deliver a substantial load of siRNA [68, 70]. Using molecular modeling and *in vivo* imaging, it was noted that compared with PEI, G2–4 flexible, triazine dendrimers (25 KDa) could form thermodynamically more stable complexes with siRNA with lesser toxicity. It is emphasized that the triazine dendrimeric siRNA delivery systems were charge-neutralized more efficiently when compared to PEI complexes, and this reduced agglomeration. The range of the hydrodynamic diameters was from 72.0–153.5 nm in case of dendriplexes in contrast to 312.8–480.0 nm for the PEI-siRNA complexes. All the evaluated dendriplexes underwent efficient endocytosis and demonstrated significantly more luciferase knockdown. However, the higher generation dendriplexes were taken up by the reticuloendothelial system owing to their increased surface charge.

The carbosilane dendrimers, another new class, has been employed for siRNA delivery [69, 71]. Researchers reported the development of amino terminated carbosilane dendrimers to protect/transport siRNA to HIV-infected lymphocytes [69]. The dendrimers could effectively bind siRNA through electrostatic interactions, and the dendrimer-siRNA complex was resistant to RNAse-mediated degradation. The dendriplexes having an N/P ratio of 2 exhibited the maximum transfection efficiency without cytotoxicity in the otherwise difficult-to-transfect peripheral blood mononuclear cells infected with HIV. The dendrimer-siRNA complex could down-regulate the GAPDH expression and decrease HIV-replication in the tested cell lines. Carbosilane dendrimers have also been employed to deliver siRNA to the post-mitotic neurons for studying the function of HIF1-α (hypoxia-inducible factor-1 alpha) in a model of chemical hypoxia-induced neurotoxicity [71]. These carbosilane dendrimers were no less effective as viral vectors with respect to the siRNA delivery and efficiency of transfection. Carbosilane dendrimers with 16 positive charges per molecule resulted in a powerful repression of many interleukins in macrophages participating in autoimmune diseases thereby suggesting a potential clinical application of these dendrimers [72].

The PPI (polypropyleneimine) dendrimers also appear to be attractive non-viral vectors for siRNA delivery [57]. By using a layer-by-layer surface modification technique, PPI dendrimer-siRNA nanoparticles were formulated wherein the nanoparticles were caged with a disulfhydryl containing crosslinker. This was followed by coating with a PEG polymer for providing stability to the nanoparticles necessary to combat the neutralizing environment in the circulation. Specific tumor targeting of these nanoparticles could be achieved by conjugating a tumor-homing peptide, i.e., LHRH (luteinizing hormone-releasing hormone) with the distal end of the PEG-chain. The disulfide bonds got reduced following cellular uptake and the nucleic acid was released into the cytosol [73]. It is appreciated that coating with a disulfhydryl linkage on the dendriplex provided stability thereby resulting in less agglomeration of the nanoparticles in the blood stream [74]. However, this may compromise the transfection efficiency because of over-stabilization of the product and hence may require the use of less stable disulfide bonds [75]. Surface functionalization with thiol linkage is useful to stabilize the nanoparticles by preventing dissociation in the presence of polyanionic competitors. The modified nanoparticles are stable in human serum for a minimum of 48 h. PEGylation imparts stability against agglomeration by bringing down the particle-protein and particle-particle interactions. Thus, for cancer targeting, the functionalization of nanoparticle surface with the LHRH peptide offers an effective strategy. The LHRH receptors are known to be over-expressed in many cancers such as prostate, ovarian, lung, colon, and breast cancer in contrast to an undetectable expression in otherwise healthy tissues [76]. The results of this *in vivo* biodistribution study suggested a high degree of accumulation of nanoparticles in tumor tissue as compared to other organs.

Other researchers have used lactose-modified, α-cyclodextrin conjugated generation 3 dendrimers (Lac-α-CDE) intended to deliver liver-specific siRNA carriers. The purpose was to treat familial amyloidotic polyneuropathy (transthyretin-related) [77, 78]. The Lac-α-CDE was condensed with an siRNA targeting the TTR (transthyretin) gene expression. This dendrimer-siRNA complex showed: (i) a potent RNAi against the TTR gene expression; (ii) an efficient endosomal escape; (iii) an efficient delivery of the siRNA complex to the cytosol; and (iv) negligible cytotoxicity. Thus, following intravenous administration, the Lac-α-CDE/siRNA complex could demonstrate a significant *in vivo* gene silencing.

Flexible PAMAM dendrimers having TEA (triethanolamine) as the core, and the branching units commencing at a distance of 10 consecutive bonds from the central amine was evaluated for siRNA delivery/gene silencing [79–82]. With a TEA core, there is a provision of flexibility compared to commercially PAMAM having ammonia/ethylenediamine core (in this case the branching begins at the central amine). Higher generation dendrimers within this family could efficiently deliver the siRNA and induce gene silencing. Further, the dendrimers with amine end groups almost fully retarded the siRNA during agarose gel electrophoresis at N/P ratios >2.5. However, no gel retardation was observed with dendrimers having terminal ester groups, suggesting a lack of condensation with siRNA. This model of TEA core PAMAM dendrimers successfully delivered the siRNA and inhibited the catalysis of ribozymes in *Candida* [82]. In another study, a TEA core PAMAM dendrimer was used to deliver HSP27-targeted siRNA inside the prostate cancer cells [80]. The dendrimer protected the siRNA from enzymatic degradation and promoted the cellular uptake of siRNA. Moreover, the siRNA also demonstrated potent as well as specific gene silencing of hsp 27 (heat shock protein 27), which is an attractive target in cases of castration-resistant prostate cancer.

7.6 SURFACE MODIFICATION TO IMPROVE EFFICACY AND MULTIFUNCTIONALITY

The surface functionalization of nanoparticulate carriers (simply 'nanocarriers') with a wide range of polymers and targeting ligands is a useful and promising approach for achieving specific desired functions [83]. Thus, the surface modification of nanocarriers with PEG, a biocompatible and hydrophilic polymer, has found extensive applications. This is because pegylation protects the nanocarrier from exposure to degrading enzymes or even opsonins in the systemic circulation [84–89]. Cleverly evading recognition and capture by the reticuloendothelial system ensures a longer half life in the systemic circulation. This is particularly essential for those nanocarriers which eventually need to accumulate in an infarcted area or a tumor through the phenomenon of EPR (enhanced permeability and retention) [84, 90, 91]. Pegylation also decreases the cytotoxicity of cationic polymers by decreasing or partially occluding the positive charges on the surface of the polycations [84, 87, 89, 92]. Higher generation PAMAM dendrimers (G4 and G5) are remarkably efficient as nucleic acid delivery vectors. Nevertheless, their hepatotoxic and hemolytic potential limit their application *in vivo* [93–95].

In other studies, PEG-modified PAMAM dendrimers have shown reduced toxicity vis-à-vis maintaining efficient drug/gene delivery [96–98]. The extent of surface pegylation affects both the attributes, i.e., the transfection efficacy and cytotoxicity. The G5 PAMAM dendrimer, linked with 10% PEG of 3.4 KDa demonstrated a 20-fold increase in the efficacy of gene transfection when compared with the unconjugated G5 PAMAM dendrimer [99]. Researchers have. demonstrated 8 mol % pegylated G5/G6 dendrimers to be the most efficient for gene silencing, in comparison to 3 different systems of pegylated G5 and G6 PAMAM dendrimers (evaluated at 4, 8, or 15% molar ratio of surface PEG) [96]. In another report, G5-PAMAM dendrimers were linked with cell penetrating TAT peptide for achieving intracellular delivery [100]. The *MDR1* gene silencing siRNA-dendrimer complexes were weak at inhibiting the gene expression. Conjugation with TAT-peptide failed to improve the delivery of the G5-PAMAM. However, the TAT-functionalized dendrimer-oligonucleotide polyplexes proved to be moderately effective in delivering the antisense in comparison to the plain dendrimer. Further, the effect of acetylating the primary amines (in G5) on dendriplex formation was also evaluated [101]. The results suggested that acetylation reduces the cytotoxicity and that nearly 20% of the primary amines could be modified besides maintaining the siRNA delivery efficiency noted for the unmodified PAMAM. A greater degree of amine neutralization led to a reduction in the cellular delivery, causing entrapment in the endosomes owing to the reduction in buffering capacity, thus compromising the gene silencing efficiency.

A novel approach for cutting down the cytotoxicity of dendrimers has been reported by a group of researchers [102–105] who evaluated internal cationic but surface neutral dendrimers for the targeted delivery of siRNA. They synthesized internal quaternized, surface acetylated PAMAM-G4 dendrimers. These dendrimers elicited remarkably low cytotoxicity when compared with the parent dendrimers having surface amino groups. Further, strong interaction of siRNA with the cationic charge distributed inside the dendrimer led to the formation of a compact nanoparticle that could potentially protect the siRNA from degradation. The shape of the PAMAM/siRNA polyplexes was spherical at a charge N/P ratio of 3. On the other hand, at the same N/P ratio, the condensation of siRNA with PAMAM-NH2 led to the synthesis of ribbon-like nanofibers [103]. The PAMAM-OH dendrimers were internally cationic due to the reaction between tertiary amines and methyl iodide. This charge was exploited for siRNA condensation. It is emphasized that the internal condensation of siRNA played a key role in controlling the morphology of the polyplexes due to the encapsulating of the siRNA in the core. Next, the quaternized PAMAM-OH G4 dendrimers (QPAMAM-OH) were functionalized with a customized moiety, a synthetic analog of the natural peptide LHRH, to achieve cancer targeting [105]. The QPAMAM-OH-LHRH successfully localized to and accumulated in the cancer cells and enhanced the cellular uptake of dendrimer-siRNA polyplexes through the interaction with the over-expressed LHRH receptors (via receptor-mediated endocytosis). There was minimal invasion of the healthy tissues thus limiting the adverse effects.

Extensive work has shown that for successful siRNA delivery, functionalized inorganic nanomaterials such as quantum dots, gold, iron oxide nanoparticles and carbon nanotubes can serve as a promising platform. It is because these materials generate a high molecular weight polymeric structure bearing surface functional groups [106–108]. The inorganic core serves to function as a space-filling material which presents the surface-associated functional groups. Thus, gold nanoparticles have been functionalized using the biodegradable glutamic acid scaffolds and the positively charged triethylenetetramine terminated dendron ligands to ensure effective electrostatic interactions with siRNA [106]. In another study, PPI (poly-propyleneimine) G5 dendrimers were cooperatively linked with SPION (super paramagnetic iron oxide nanoparticles) and siRNA for developing a tumor-targeting drug delivery system. It served dual purpose, i.e., simultaneous delivery of siRNA and the MRI contrast agent specifically to tumor cells [107]. PEG-stabilized dendrimer-siRNA-SPION particles with the distal ends coupled with the LHRH (the tumor homing peptide) could deliver the siRNA specifically to the tumor cells. The best part was, the protocol permitted the monitoring of the treatment outcome by using concurrent imaging techniques.

In another study, G4 PAMAM dendrimers were lipid-modified for co-delivering siRNA-drug [109]. The researchers prepared a triple block co-polymer system by linking PAMAM G4 dendrimers with PEG-PE (PEG-1,2-dioleoyl-sn-glycero-3-phosphoethanolamine). The lipid unit provided an optimum hydrophobicity besides a compatible cellular interaction ensuring an enhanced cell penetration. Subsequently, a mixed micellar system was developed by employing G4-D-PEG-PE and PEG-PE polymer (molar ratio 1:1). The systems formed stable complexes with siRNA and demonstrated good serum stability. A significantly higher cellular uptake was observed for siRNA and resulted in better targeted GFP (green fluorescence protein) down-regulation when compared with PAMAM G4 dendrimer. The mixed micellar system core efficiently loaded doxorubicin, an anticancer drug, while the PEG-PE linked G4-D condensed the siRNA. The modified dendrimer demonstrated higher efficiency for siRNA delivery compared to G(4)-D and the mixed micellar system. Thus, the latter appears to be a promising candidate for the co-delivery of siRNA/drug. Confocal microscopy image results of the study demonstrated beyond doubt that the lipid modification augments the cellular association of the nano-system without reducing the dendrimers' ability for siRNA condensation. Further, such a mixed micellar system is advantageous for the dual ability to concurrently carry and effectively co-deliver drug/siRNA.

7.7 NOW AND NEXT?

Overall, dendrimer-mediated delivery of therapeutic nucleic acids including siRNA definitely is to be considered a promising approach. The unique features of dendrimeric systems including the ease of surface functionalization, enables the crafting of truly versatile nanodevices for drug delivery applications. However, careful addressing of the recognized current limitations that include non-specific cytotoxicity observed with higher generation-dendrimers, release kinetics and fast clearance would open up new vistas for dendrimers as state of the art gene delivery carriers. Looking at the toxicological issues and other demerits of the viral approach, the non-viral vectors are increasingly and rightly being considered as suitable alternatives. Further, among the versatile non-viral vectors for siRNA delivery, PAMAM dendrimers are the most extensively investigated largely due to the ease of synthesis/modifications as well as the commercial availability. Three basic means for optimizing the dendrimer structure have emerged in order to lower the toxicity vis-à-vis enhancing the efficiency of delivery [85, 110]: (i) synthesis of novel dendrimer structure or resorting to new core unit; (ii) functionalizing the interior/exterior of dendrimers; and (iii) using other biocompatible and bioactive molecules that can form stable complexes with dendrimers. Because various cell lines have variable requirements for a particular delivery system, one needs to be careful while evaluating dendrimer-based siRNA delivery, choosing the appropriate dendrimer structure, optimizing the method as well as the delivery conditions. Each of these, of course, would depend on the type of target gene and the type of cells bearing those genes.

KEYWORDS

- **carbosilane**
- **delivery vectors**
- **dendrimer multifunctionality**
- **dicer**
- **interfering nanoparticles**
- **siRNA**

REFERENCES

1. Fire, A., Xu, S., Montgomery, M. K., Kostas, S. A., Driver, S. E., & Mello, C. (1998). "Potent and specific genetic interference by double-stranded RNA in *Caenorhabditis elegans*," *Nature, 391*(6669), 806–811.
2. Hammond, S. M., Boettcher, S., Caudy, A. A., Kobayashi, R., & Hannon, G. J. (2001). "Argonaute, a link between genetic and biochemical analyses of RNAi," *Science, 293*(5532), 1146–1150.
3. Aagaard, L., & Rossi, J. J. (2007). "RNAi therapeutics: principles, prospects and challenges," *Advanced Drug Delivery Reviews, 59*(2–3), 75–86.
4. Lal, H., & Pandey, R. *Textbook of Biochemistry*, 2nd Edition, CBS Publishers, New Delhi, India, 2011, pp. 652. ISBN: 978-81-239-2039-9.
5. Joshi, M., Sodhi, K. S., Pandey, R., Singh, J., & Goyal, S. (2014). RNA aptamer technology. *Indo American J Pharm Res 4*(9), 3676–368.
6. Wang, J., Lu, Z., Wientjes, M. G., & Au, S. J. L. (2010). "Delivery of siRNA therapeutics: barriers and carriers," *AAPS Journal, 12*(4), 492–503.
7. Zhou, J., Wu, J., Hafdi, N., Behr, J.-P., Erbacher, P., & Peng, L. (2006). "PAMAM dendrimers for efficient siRNA delivery and potent gene silencing," *Chemical Communications. 22*, 2362–2364.
8. Liu, X.-X., Rocchi, P., Qu, F.-Q. et al. (2009). "PAMAM dendrimers mediate siRNA delivery to target Hsp27 and produce potent antiproliferative effects on prostate cancer cells," *Chem Med Chem, 4*(8), 1302–1310.
9. Zhou, J., Neff, C. P., Liu, X. et al. (2011). "Systemic administration of combinatorial dsiRNAs via nanoparticles efficiently suppresses HIV-1 infection in humanized mice," *Molecular Therapy, 19*(12), 2228–2238.
10. Liu, X., Liu, C., Laurini, E. et al. (2012). "Efficient delivery of sticky siRNA and potent gene silencing in a prostate cancer model using a generation 5 triethanolamine-core PAMAM dendrimer," *Molecular Pharmaceutics, 9*(3), 470–481.
11. Yu, T. Z., Liu, X. X., Bolcato-Bellemin, A. L. et al. (2012). "An amphiphilic dendrimer for effective delivery of small interfering RNA and gene silencing in vitro and in vivo," *Angewandte Chemie International Edition, 51*(34), 8478–8484.
12. Taratula, O., Garbuzenko, O. B., Kirkpatrick, P. et al. (2009). "Surface-engineered targeted PPI dendrimer for efficient intracellular and intratumoral siRNA delivery," *Journal of Controlled Release, 140*(3), 284–293.
13. Chen, A. M., Taratula, O., Wei, D. et al. (2010). "Labile catalytic packaging of DNA/siRNA: control of gold nanoparticles "out" of DNA/siRNA complexes," *ACS Nano, 4*(7), 3679–3688.
14. Taratula, O., Savla, R., He, H., & Minko, T. (2011). "Poly(propylene-imine) dendrimers as potential siRNA delivery nanocarrier: from structure to function," *International Journal of Nanotechnology, 8*(1–2), 36–52.
15. Taratula, O., Garbuzenko, O., Savla, R., Wang, Y. A., He, H., & Minko, T. (2011). "Multifunctional nanomedicine platform for cancer specific delivery of siRNA by superparamagnetic iron oxide nanoparticles-dendrimer complexes," *Current Drug Delivery, 8*(1), 59–69.

16. Jimenez, J. L., Gomez, R., Briz, V. et al. (2012). "Carbosilane dendrimers as carriers of siRNA," Journal of Drug Delivery Science and Technology, 22(1), 75–82.
17. Weber, N., Ortega, P., Clemente, M. I. et al. (2008). "Characterization of carbosilane dendrimers as effective carriers of siRNA to HIV-infected lymphocytes," *Journal of Controlled Release, 132*(1), 55–64.
18. Gras, R., Almonacid, L., Ortega, P. et al. (2009). "Changes in gene expression pattern of human primary macrophages induced by carbosilane dendrimer 2G-NN16," *Pharmaceutical Research. 26*(3), 577–586.
19. Inoue, Y., Kurihara, R., Tsuchida, A. et al. (2008). "Efficient delivery of siRNA using dendritic poly(L-lysine) for loss-of-function analysis," *Journal of Controlled Release, 126*(1), 59–66.
20. Watanabe, K., Harada-Shiba, M., A. Suzuki et al. (2009). "In vivo siRNA delivery with dendritic poly(L-lysine) for the treatment of hypercholesterolemia," *Molecular BioSystems, 5*(11), 1306–1310.
21. Kaneshiro, T. L. & Lu, Z.-R. (2009). "Targeted intracellular codelivery of chemotherapeutics and nucleic acid with a well-defined dendrimer-based nanoglobular carrier," Biomaterials, 30(29), 5660–5666.
22. Merkel, O. M., Mintzer, M. A., D. Librizzi et al. (2010). "Triazine dendrimers as nonviral vectors for in vitro and in vivo RNAi: the effects of peripheral groups and core structure on biological activity," *Molecular Pharmaceutics, 7*(4), 969–983.
23. Pavan, G. M., Mintzer, M. A., Simanek, E. E., Merkel, O. M., Kissel, T., & Danani, A. (2010). "Computational insights into the inter-actions between DNA and siRNA with "rigid" and "flexible" triazine dendrimers," *Biomacromolecules, 11*(3), 721–730.
24. Merkel, O. M., Zheng, M., M. A. Mintzer et al. (2011). "Molecular modeling and in vivo imaging can identify successful flexible triazine dendrimer-based siRNA delivery systems," *Journal of Controlled Release, 153*(1), 23–33.
25. Ofek, P., Fischer, W., Calderon, M., Haag, R., & Satchi-Fainaro, R. (2010). "In vivo delivery of small interfering RNA to tumors and their vasculature by novel dendritic nanocarriers," *FASEB Journal, 24*(9), 3122–3134.
26. Fischer, W., Calderon, M., Schulz, A., Weber, M., & Haag, R. (2010). "Dendritic polyglycerols with oligoamine shells show low toxicity and high siRNA transfection efficiency in vitro," *Bioconjugate Chemistry, 21*(10), 1744–1752.
27. Malhotra, S., Bauer, H., Tschiche, A. et al. (2012). "Glycine-terminated dendritic amphiphiles for nonviral gene delivery," *Biomacromolecules, 13*(10), 3087–3098.
28. Herrero, M. A., Toma, F. M., Al-Jamal, K. T. et al. (2009). "Synthesis and characterization of a carbon nanotube-dendron series for efficient siRNA delivery," *Journal of the American Chemical Society, 131(*28), 9843–9848.
29. McCarroll, J., Baigude, H., C.-Yang, S., & Rana, T. M. (2010). "Nanotubes functionalized with lipids and natural amino acid dendrimers: a new strategy to create nanomaterials for delivering systemic RNAi," *Bioconjugate Chemistry, 21*(1), 56– 63.
30. Guerra, J., Herrero, M. A., Carrion, B. et al., "Carbon nanohorns functionalized with polyamidoamine dendrimers as efficient biocarrier materials for gene therapy," *Carbon, 50*(8), 2832–2844. (2012).

31. Baigude, H., McCarroll, J., C.-Yang, S., Swain, P. M., & Rana, T. M. (2007). "Design and creation of new nanomaterials for therapeutic RNAi," *ACS Chemical Biology, 2*(4), 237–241.
32. Menuel, S., Fontanay, S., Clarot, I., Duval, R. E., Diez, L., & Marsura, A. (2008). "Synthesis and complexation ability of a novel bis-(guanidinium)-tetrakis-(-cyclodextrin) dendrimeric tetra-pod as a potential gene delivery (DNA and siRNA) system. Study of cellular siRNA transfection," *Bioconjugate Chemistry, 19*(12), 2357–2362.
33. Agrawal, A., D.-Min, H., N. Singh et al. (2009). "Functional delivery of siRNA in mice using dendriworms," *ACS Nano, 3*(9), 2495–2504.
34. Briz, V., Serramia, M. J., Madrid, R. et al. (2012). "Validation of a generation 4 phosphorus-containing polycationic dendrimer for gene delivery against HIV-1," *Current Medicinal Chemistry, 19*(29), 5044–5051.
35. Aagaard, L., & Rossi, J. J. (2007). Rnai therapeutics: Principles, prospects and challenges. *Adv. Drug Deliver. Rev. 59*, 75–86.
36. Elbashir, S. M., Harborth, J., Lendeckel, W., Yalcin, A., Weber, K., & Tuschl, T. (2001). Duplexes of 21-nucleotide rnas mediate rna interference in cultured mammalian cells. *Nature 411*, 494–498.
37. Daka, A., & Peer, D. (2012). RNAi-based nanomedicines for targeted personalized therapy. *Adv. Drug Deliver Rev. 64*, 1508–1521.
38. Potti, A., Schilsky, R. L., & Nevins, J. R. (2010). Refocusing the war on cancer: The critical role of personalized treatment. *Sci. Transl. Med. 2*, 2813.
39. Zimmermann, T. S., Lee, A. C., Akinc, A., Bramlage, B., Bumcrot, D., Fedoruk, M. N., Harborth, J., Heyes, J. A., Jeffs, L. B., John, M., et al. (2006). RNAi-mediated gene silencing in non-human primates. *Nature 441*, 111–114.
40. Dorn, G., Patel, S., Wotherspoon, G., Hemmings-Mieszczak, M., Barclay, J., Natt, F. J., Martin, P., Bevan, S., Fox, A., Ganju, P., et al. (2004). siRNA relieves chronic neuropathic pain. *Nucleic Acids Res. 32*, e49.
41. Shen, J., Samul, R., Silva, R. L., Akiyama, H., Liu, H., Saishin, Y., Hackett, S. F., Zinnen, S., Kossen, K., Fosnaugh, K., et al. (2006). Suppression of ocular neovascularization with siRNA targeting VEGF receptor 1. *Gene Ther. 13*, 225–234.
42. Behlke, M. A. (2008). Chemical modification of siRNAs for in vivo use. *Oligonucleotides 18*, 305–319.
43. Chiu, Y. L., & Rana, T. M. (2003). siRNA function in RNAi: A chemical modification analysis. *RNA* 9, 1034–1048.
44. Elmen, J., Thonberg, H., Ljungberg, K., Frieden, M., Westergaard, M., Xu, Y., Wahren, B., Liang, Z., Orum, H., Koch, T., et al. (2005). Locked nucleic acid (LNA) mediated improvements in siRNA stability and functionality. *Nucleic Acids Res. 33*, 439–447.
45. Allerson, C. R., Sioufi, N., Jarres, R., Prakash, T. P., Naik, N., Berdeja, A., Wanders, L., Griffey, R. H., Swayze, E. E., & Bhat, B. (2005). Fully 2'-modified oligonucleotide duplexes with improved in vitro potency and stability compared to unmodified small interfering RNA. *J. Med. Chem. 48*, 901–904.
46. Braasch, D. A., Jensen, S., Liu, Y., Kaur, K., Arar, K., White, M. A., & Corey, D. R. (2003). RNA interference in mammalian cells by chemically-modified RNA. *Biochemistry 42*, 7967–7975.

47. Morrissey, D. V., Blanchard, K., Shaw, L., Jensen, K., Lockridge, J. A., Dickinson, B., McSwiggen, J. A., Vargeese, C., Bowman, K., Shaffer, C. S., et al. (2005). Activity of stabilized short interfering RNA in a mouse model of hepatitis b virus replication. *Hepatology 41*, 1349–1356.
48. Prakash, T. P., Allerson, C. R., Dande, P., Vickers, T. A., Sioufi, N., Jarres, R., Baker, B. F., Swayze, E. E., Griffey, R. H., & Bhat, B. (2005). Positional effect of chemical modifications on short interference RNA activity in mammalian cells. *J. Med. Chem. 48*, 4247–4253.
49. Choi, H. S., Liu, W., Misra, P., Tanaka, E., Zimmer, J. P., Itty Ipe, B., Bawendi, M. G., & Frangioni, J. V. (2007). Renal clearance of quantum dots. *Nat. Biotechnol. 25*, 1165–1170.
50. Van de Water, F. M., Boerman, O. C., Wouterse, A. C., Peters, J. G., Russel, F. G., & Masereeuw, R. (2006). Intravenously administered short interfering RNA accumulates in the kidney and selectively suppresses gene function in renal proximal tubules. *Drug Metab. Dispos. 34*, 1393–1397.
51. Jeong, J. H., Park, T. G., & Kim, S. H. (2011). Self-assembled and nanostructured siRNA delivery systems. *Pharm. Res. 28*, 2072–2085.
52. Sledz, C. A., & Williams, B. R. (2005). RNA interference in biology and disease. *Blood 106*, 787–794.
53. Kariko, K., Bhuyan, P., Capodici, J., & Weissman, D. (2004). Small interfering RNAs mediate sequence-independent gene suppression and induce immune activation by signaling through toll-like receptor 3. *J. Immunol. 172*, 6545–6549.
54. Jackson, A. L., Bartz, S. R., Schelter, J., Kobayashi, S. V., Burchard, J., Mao, M., Li, B., Cavet, G., & Linsley, P. S. (2003). Expression profiling reveals off-target gene regulation by RNAi. *Nat. Biotechnol. 21*, 635–637.
55. Merkel, O. M., Mintzer, M. A., Librizzi, D., Samsonova, O., Dicke, T., Sproat, B., Garn, H., Barth, P. J., Simanek, E. E., & Kissel, T. (2010). Triazine dendrimers as nonviral vectors for in vitro and in vivo RNAi: The effects of peripheral groups and core structure on biological activity. *Mol. Pharm. 7*, 969–983.
56. Kebbekus, P., Draper, D. E., & Hagerman, P. (1995). Persistence length of RNA. *Biochemistry 34*, 4354–4357.
57. Taratula, O., Garbuzenko, O. B., Kirkpatrick, P., Pandya, I., Savla, R., Pozharov, V. P., He, H., & Minko, T. (2009). Surface-engineered targeted PPI dendrimer for efficient intracellular and intratumoral siRNA delivery. *J. Control. Release 140*, 284–293.
58. Spagnou, S., Miller, A. D., & Keller, M. (2004). Lipidic carriers of siRNA: Differences in the formulation, cellular uptake, and delivery with plasmid DNA. *Biochemistry 43*, 13348–13356.
59. Gary, D. J., Puri, N., & Won, Y. Y. (2007). Polymer-based siRNA delivery: Perspectives on the fundamental and phenomenological distinctions from polymer-based DNA delivery. *J. Control. Release 121*, 64–73.
60. Haensler, J., & Szoka, F. C., Jr. (1993). Polyamidoamine cascade polymers mediate efficient transfection of cells in culture. *Bioconjug. Chem. 4*, 372–379.
61. Kim, T. I., Seo, H. J., Choi, J. S., Jang, H. S., Baek, J. U., Kim, K., & Park, J. S. (2004). PAMAM-PEG-PAMAM: Novel triblock copolymer as a biocompatible and efficient gene delivery carrier. *Biomacromolecules 5*, 2487–2492.

62. Schatzlein, A. G., Zinselmeyer, B. H., Elouzi, A., Dufes, C., Chim, Y. T., Roberts, C. J., Davies, M. C., Munro, A., Gray, A. I., & Uchegbu, I. F. (2005). Preferential liver gene expression with polypropylenimine dendrimers. *J. Control. Release 101,* 247–258.
63. Forrest, M. L., Gabrielson, N., & Pack, D. W. (2005). Cyclodextrin-polyethylenimine conjugates for targeted in vitro gene delivery. *Biotechnol. Bioeng. 89*, 416–423.
64. Richardson, S. C., Pattrick, N. G., Man, Y. K., Ferruti, P., & Duncan, R. (2001). Poly(amidoamine)s as potential nonviral vectors: Ability to form interpolyelectrolyte complexes and to mediate transfection in vitro. *Biomacromolecules 2*, 1023–1028.
65. Jensen, L. B., Pavan, G. M., Kasimova, M. R., Rutherford, S., Danani, A., Nielsen, H. M., & Foged, C. (2011). Elucidating the molecular mechanism of PAMAM-siRNA dendriplex self-assembly: Effect of dendrimer charge density. *Int. J. Pharm. 416*, 410–418.
66. Posadas, I., Guerra, F. J., & Cena, V. (2010). Nonviral vectors for the delivery of small interfering RNAs to the CNS. *Nanomedicine 5,* 1219–1236.
67. Ofek, P., Fischer, W., Calderon, M., Haag, R., & Satchi-Fainaro, R. (2010). In vivo delivery of small interfering RNA to tumors and their vasculature by novel dendritic nanocarriers. *FASEB J. 24*, 3122–3134.
68. Mintzer, M. A., Merkel, O. M., Kissel, T., & Simanek, E. E. (2009). Polycationic triazine-based dendrimers: Effect of peripheral groups on transfection efficiency. *New J. Chem. 33*, 1918–1925.
69. Weber, N., Ortega, P., Clemente, M. I., Shcharbin, D., Bryszewska, M., de la Mata, F. J., Gomez, R., & Munoz-Fernandez, M. A. (2008). Characterization of carbosilane dendrimers as effective carriers of siRNA to HIV-infected lymphocytes. *J. Control. Release 132*, 55–64.
70. Merkel, O. M., Mintzer, M. A., Sitterberg, J., Bakowsky, U., Simanek, E. E., Kissel, T. (2009). Triazine dendrimers as nonviral gene delivery systems: Effects of molecular structure on biological activity. *Bioconjug. Chem. 20*, 1799–1806.
71. Posadas, I., Lopez-Hernandez, B., Clemente, M. I., Jimenez, J. L., Ortega, P., de la Mata, J., Gomez, R., Munoz-Fernandez, M. A., & Cena, V. (2009). Highly efficient transfection of rat cortical neurons using carbosilane dendrimers unveils a neuroprotective role for hif-1alpha in early chemical hypoxia-mediated neurotoxicity. *Pharm. Res. 26*, 1181–1191.
72. Gras, R., Almonacid, L., Ortega, P., Serramia, M. J., Gomez, R., de la Mata, F. J., Lopez-Fernandez, L. A., & Munoz-Fernandez, M. A. (2009). Changes in gene expression pattern of human primary macrophages induced by carbosilane dendrimer 2g-nn16. *Pharm. Res. 26*, 577–586.
73. Ooya, T., Lee, J., & Park, K. (2004). Hydrotropic dendrimers of generations 4 and 5: Synthesis, characterization, and hydrotropic solubilization of paclitaxel. *Bioconjug. Chem. 15*, 1221–1229.
74. Trubetskoy, V. S., Loomis, A., Slattum, P. M., Hagstrom, J. E., Budker, V. G., & Wolff, J. A. (1999). Caged DNA does not aggregate in high ionic strength solutions. *Bioconjugate Chem 10*, 624–628.
75. Miyata, K., Kakizawa, Y., Nishiyama, N., Harada, A., Yamasaki, Y., Koyama, H., & Kataoka, K. (2004). Block catiomer polyplexes with regulated densities of charge

and disulfide cross-linking directed to enhance gene expression. *J. Am. Chem. Soc. 126*, 2355–2361.

76. Dharap, S. S., Wang, Y., Chandna, P., Khandare, J. J., Qiu, B., Gunaseelan, S., Sinko, P. J., Stein, S., Farmanfarmaian, A., & Minko, T. (2005). Tumor-specific targeting of an anticancer drug delivery system by LHRH peptide. *Proc. Natl. Acad. Sci. USA 102*, 12962–12967.
77. Hayashi, Y., Mori, Y., Yamashita, S., Motoyama, K., Higashi, T., Jono, H., Ando, Y., & Arima, H. (2012). Potential use of lactosylated dendrimer (G3)/α-cyclodextrin conjugates as hepatocyte-specific siRNA carriers for the treatment of familial amyloidotic polyneuropathy. *Mol. Pharm. 9*, 1645–1653.
78. Hayashi, Y., Mori, Y., Higashi, T., Motoyama, K., Jono, H., Sah, D. W., Ando, Y., & Arima, H. (2012). Systemic delivery of transthyretin siRNA mediated by lactosylated dendrimer/α-cyclodextrin conjugates into hepatocyte for familial amyloidotic polyneuropathy therapy. *Amyloid 19*(Suppl 1), 47–49.
79. Zhou, J., Wu, J., Hafdi, N., Behr, J. P., Erbacher, P., & Peng, L. (2006). PAMAM dendrimers for efficient siRNA delivery and potent gene silencing. *Chem. Commun. 22*, 2362–2364.
80. Liu, X. X., Rocchi, P., Qu, F. Q., Zheng, S. Q., Liang, Z. C., Gleave, M., Iovanna, J., & Peng, L. (2009). PAMAM dendrimers mediate siRNA delivery to target hsp27 and produce potent antiproliferative effects on prostate cancer cells. *Chem. Med. Chem. 4*, 1302–1310.
81. Shen, X. C., Zhou, J., Liu, X., Wu, J., Qu, F., Zhang, Z. L., Pang, D. W., Quelever, G., Zhang, C. C., & Peng, L. (2007). Importance of size-to-charge ratio in construction of stable and uniform nanoscale RNA/dendrimer complexes. *Org. Biomol. Chem. 5*, 3674–3681.
82. Wu, J., Zhou, J., Qu, F., Bao, P., Zhang, Y., & Peng, L. (2005). Polycationic dendrimers interact with RNA molecules: Polyamine dendrimers inhibit the catalytic activity of candida ribozymes. *Chem. Commun. 41*, 313–315.
83. Torchilin, V. P. (2006). Multifunctional nanocarriers. *Adv. Drug Deliver. Rev. 58*, 1532–1555.
84. Lee, M., Kim, S. W. (2005). Polyethylene glycol-conjugated copolymers for plasmid DNA delivery. *Pharm. Res. 22*, 1–10.
85. Biswas, S., & Torchilin, V. P. (2013). Dendrimers for siRNA Delivery. *Pharmaceuticals 6*, 161–183.
86. Bhadra, D., Bhadra, S., J& ain, N. K. (2005). Pegylated lysine based copolymeric dendritic micelles for solubilization and delivery of artemether. *J. Pharm. Sci. 8*, 467–482.
87. Choi, Y. H., Liu, F., Kim, J. S., Choi, Y. K., Park, J. S., & Kim, S. W. (1998). Polyethylene glycol-grafted poly-l-lysine as polymeric gene carrier. *J. Control. Release 54*, 39–48.
88. Morato, R. G., Bueno, M. G., Malmheister, P., Verreschi, I. T., & Barnabe, R. C. (2004). Changes in the fecal concentrations of cortisol and androgen metabolites in captive male jaguars (*Panthera onca*) in response to stress. *Braz. J. Med. Biol. Res. 37*, 1903–1907.
89. Kursa, M., Walker, G. F., Roessler, V., Ogris, M., Roedl, W., Kircheis, R., & Wagner, E. (2003). Novel shielded transferrin-polyethylene glycol-polyethylenimine/

DNA complexes for systemic tumor-targeted gene transfer. *Bioconjug. Chem. 14,* 222–231.
90. Torchilin, V. P., Omelyanenko, V. G., Papisov, M. I., Bogdanov, A. A., Jr., Trubetskoy, V. S., Herron, J. N., & Gentry, C. A. (1994). Poly(ethylene glycol) on the liposome surface: On the mechanism of polymer-coated liposome longevity. *Biochem. Biophys. Acta 1195*, 11–20.
91. Maeda, H., Sawa, T., & Konno, T. (2001). Mechanism of tumor-targeted delivery of macromolecular drugs, including the EPR effect in solid tumor and clinical overview of the prototype polymeric drug smancs. *J. Control. Release 74,* 47–61.
92. Bikram, M., Ahn, C.-H., Chae, S. Y., Lee, M., Yockman, J. W., & Kim, S. W. (2004). Biodegradable poly (ethylene glycol)-co-poly(l-lysine)-g-histidine multiblock copolymers for nonviral gene delivery. *Macromolecules 37,* 1903–1916.
93. Roberts, J. C., Bhalgat, M. K., & Zera, R. T. (1996). Preliminary biological evaluation of polyamidoamine (pamam) starburst dendrimers. *J. Biomed. Mater. Res. 30,* 53–65.
94. Malik, N., Wiwattanapatapee, R., Klopsch, R., Lorenz, K., Frey, H., Weener, J. W., Meijer, E. W., Paulus, W., & Duncan, R. (2000). Dendrimers: Relationship between structure and biocompatibility in vitro, and preliminary studies on the biodistribution of 125I-labeled polyamidoamine dendrimers in vivo. *J. Control. Release 65,* 133–148.
95. Chen, H. T., Neerman, M. F., Parrish, A. R., & Simanek, E. E. (2004). Cytotoxicity, hemolysis, and acute in vivo toxicity of dendrimers based on melamine, candidate vehicles for drug delivery. *J. Am. Chem. Soc. 126*, 10044–10048.
96. Qi, R., Gao, Y., Tang, Y., He, R. R., Liu, T. L., He, Y., Sun, S., Li, B. Y., Li, Y. B., & Liu, G. (2009). Peg-conjugated PAMAM dendrimers mediate efficient intramuscular gene expression. *AAPS J. 11*, 395–405.
97. Liu, M., Kono, K., & Fréchet, J. M. J. (1999). Water-soluble dendrimer–poly (ethylene glycol) star like conjugates as potential drug carriers. *J. Polymer Sci. Part A 37*, 3492–3503.
98. Tang, Y., Li, Y. B., Wang, B., Lin, R. Y., van Dongen, M., Zurcher, D. M., Gu, X. Y., Banaszak Holl, M. M., Liu, G., & Qi, R. (2012). Efficient in vitro siRNA delivery and intramuscular gene silencing using peg-modified PAMAM dendrimers. *Mol. Pharm. 9*, 1812–1821.
99. Luo, D., Haverstick, K., Belcheva, N., Han, E., & Saltzman, W. M. (2002). Poly(ethylene glycol)-conjugated PAMAM dendrimer for biocompatible, high-efficiency DNA delivery. *Macromolecules 35,* 3456–3462.
100. Kang, H., DeLong, R., Fisher, M. H., & Juliano, R. L. (2005). Tat-conjugated PAMAM dendrimers as delivery agents for antisense and siRNA oligonucleotides. *Pharm. Res. 22*, 2099–2106.
101. Waite, C. L., Sparks, S. M., Uhrich, K. E., & Roth, C. M. (2009). Acetylation of PAMAM dendrimers for cellular delivery of siRNA. *BMC Biotechnol. 9*, 38.
102. Minko, T., Patil, M. L., Zhang, M., Khandare, J. J., Saad, M., Chandna, P., & Taratula, O. (2010). LHRH-targeted nanoparticles for cancer therapeutics. *Methods Mol. Biol. 624*, 281–294.

103. Patil, M. L., Zhang, M., Betigeri, S., Taratula, O., He, H., & Minko, T. (2008). Surface-modified and internally cationic polyamidoamine dendrimers for efficient siRNA delivery. *Bioconjugate Chem. 19,* 1396–1403.
104. Patil, M. L., Zhang, M., & Minko, T. (2011). Multifunctional triblock nanocarrier (pamam-peg-pll) for the efficient intracellular siRNA delivery and gene silencing. *ACS Nano 5*, 1877–1887.
105. Patil, M. L., Zhang, M., Taratula, O., Garbuzenko, O. B., He, H., & Minko, T. (2009). Internally cationic polyamidoamine pamam-oh dendrimers for siRNA delivery: Effect of the degree of quaternization and cancer targeting. *Biomacromolecules 10,* 258–266.
106. Kim, S. T., Chompoosor, A., Yeh, Y.-C., Agasti, S. S., Solfiell, D. J., & Rotello, V. M. (2012). Dendronized gold nanoparticles for siRNA delivery. *Small 8*, 3253–3256.
107. Taratula, O., Garbuzenko, O., Savla, R., Wang, Y. A., He, H., & Minko, T. (2011). Multifunctional nanomedicine platform for cancer specific delivery of siRNA by superparamagnetic iron oxide nanoparticles-dendrimer complexes. *Curr. Drug Deliver. 8*, 59–69.
108. Mishr, R. K., Dhillon, J., Pandey, R., Sodhi, K. S., & Singh, J. (2014). Quantum dots in medical science. *J Pharm Biomed Sci 4*(12), 1042–1049.
109. Biswas, S., Deshpande, P. P., Navarro, G., Dodwadkar, N. S., & Torchilin, V. P. (2013). Lipid modified triblock PAMAM-based nanocarriers for siRNA drug co-delivery. *Biomaterials 34*, 1289–1301.
110. Jiangyu Wu, Weizhe Huang, & Ziying He (2013). Dendrimers as carriers for siRNA delivery and gene silencing: A review. *Sci World J.* Article ID 630654. http://dx.doi.org/10.1155/2013/630654.

CHAPTER 8

DENDRIMERS IN DIAGNOSTICS MAGNETIC RESONANCE IMAGING (MRI)

ZAHOOR AHMAD PARRY, PhD, and RAJESH PANDEY, MD

CONTENTS

8.1 INTRODUCTION

In medical science, magnetism has an interesting, long history. In the 10th century A.D. Avicenna, an Egyptian physician cum philosopher, prescribed a grain of magnetite which was dissolved in milk for the treatment of accidental ingestion of rust. He reasoned that magnetite will render the toxic iron inert by attracting it and promoting its elimination through the intestine [1]. Nearly one thousand years later, in 1977, a small machine called 'Indomitable' was constructed. This machine used magnetism and after toiling for about 5 hours, could produce one image. This event changed the perspective of medical diagnostics and in fact modern medicine as a whole [2]. While looking at the homemade superconducting magnet prepared from 30 miles of niobium-titanium wire currently archived at its rightful place, i.e., the Smithsonian Institution, it is difficult to comprehend how in just 30 years magnetic resonance imaging (MRI) has evolved from a crude human scanner to where doctors can now routinely order MRIs owing to its exquisite anatomical resolution.

In 1994, the first dendrimer-based contrast agents were reported. The application of dendrimeric- scaffolds for MR contrast agents has generated plenty of interest [3]. The tremendous potential of this class of molecules for the development of diagnostic contrast agents lies in the fact that their synthesis allows the well-controlled, judicious use of space in 3-dimensions which may vary suitably as a function of their size, shape and functional groups [4]. The initial report of dendrimer-based MR contrast agents outlined the conjugation of G2/G6 PAMAM-Am dendrimers and Gd^{3+}-1B4M [3]. Owing to their large molecular weight (and therefore, a large molecular tumbling rate represented as τR), these agents demonstrated very high longitudinal relaxivities. Speaking in terms of molar relaxivity, G6 dendrimer was found to be almost 6 times that of Gd^{3+}-DTPA itself. Due to the potential utility of these compounds, a better-refined synthetic approach was reported of late that involved the use of non-aqueous conjugation chemistry [5]. Researchers have reported the synthesis of G0, G2 and G4 PPI dendrimers in series, conjugated with Gd^{3+}-DTPA and observed that molecular as well as ionic relaxivities could be increased as a function of generation number [6]. Likewise, other workers have reported that the estimated relaxivities of PAMAM-Am-DO3A dendrimers varying from G2–G5, as well as the higher generations of PAMAM-EDA dendrimers linked with Gd^{3+}-DOTA, varying from G5–G10, increased with increasing molecular weight [7, 8]. However, researchers also observed that after G7, the molar relaxivities achieved a saturation limit. Other workers performed a series of varying pressure and temperature ^{17}O-NMR experiments with Gd^{3+}-DO3A labeled PAMAM-Am G3–G5 dendrimers in order to study the influence of rotational dynamics and water exchange on the relaxivity of these entities [9]. Their results suggested that while τR increases roughly by a fourth with every increase in generation, the water exchange rate constants k_{ex} remained the same for all the studied systems, thereby giving up any theoretical increment in molar relaxivity. Thus, these systems bear rotational correlation times that are long enough for the rate of water exchange to influence the net relaxivity of the dendrimer. In addition, any further improvements would mean not merely increasing the molecular weight, but also to design chelating systems which would promote the dissociation of water molecules linked to the paramagnetic Gd^{3+} ion. Moreover, their results suggested that conjugation of the macrocyclic chelate with the dendrimer does not affect the rate of exchange of water at the metal center. This indicates that the k_{ex} value determined for a particular monomeric chelate would apply to any potential dendrimeric conjugate developed in future. To this effect, other workers synthesized a series of G5 – G9 PAMAM-EDA dendrimers that were conjugated with a novel ligand called EPTPA (ethylene propylene triamine penta-acetic acid) [10]. ^{17}O-NMR experiments have demontrated that Gd^{3+}-EPTPA bears a water exchange rate that is nearly 10-fold greater than that observed with Gd^{3+}-DTPA [11]. This could be attributed to steric crowding around Gd^{3+}, which may accelerate the dissociation of the bound solvent molecules. The relaxivities of the systems went up from G5 to G7 (at 37°C, 30 MHz) which demonstrated the beneficial effect of working with chelates having more rapid water exchange rates. However, this trend was noted to decrease upon reaching G9. Again, the relaxivity measurements at varying pH suggested that the protonation of tertiary amines in the dendrimer resulted in a more open and rigid structure thereby improving the relaxivity. Hence, it was reasoned that even with more rapid water exchange kinetics, the net relaxivity of higher generation dendrimers is influenced by the internal motion as well. Similarly, other workers have reported an improved relaxivity of a PAMAM-EDA-G5 dendrimer that was conjugated with a DTPA-based chelate having one phosphinate group, DTTAP, and suggested the merits of faster water exchange besides the role played by the secondary hydration sphere [12].

Some novel ideas include the preparation of G0 and G2 PPI dendrimers that are functionalized with Yb^{3+}-DOTAM as a PARACEST (pH-sensitive, paramagnetic chemical exchange saturation transfer) agent. In this, the maximum effect was noted with falling pH from the mononuclear

chelate to dendrimer G2 [13]. The PARACEST agents have gained greater interest in molecular imaging because the paramagnetic ions induce major shifts in the resonance of the nearby nuclei that can be visualized at will simply by the correct choice of the irradiation frequency [14]. The various chelates that have been used to functionalize dendrimers are shown in Figure 8.1.

Finally, a series of Gd^{3+}-chelated core branched alcohol dendrimers have been synthesized through a convergent approach [15]. Unidirectional, amino-substituted compounds of increasing length/branching were conjugated with a Gd^{3+} chelate bearing a DOTA-like ligand (having peripheral carboxylate groups). By placing the Gd^{3+} ion at the center of the macromolecular structure, the researchers could effectively couple the local movement of Gd^{3+}-OH2 vector with that of the rotation of the complete assembly. This resulted in an increased relaxivity (Figure 8.2) [16]. The ^{17}O-NMR calculations indeed showed a greater length as well as degree of arborol branching of the entire complex when compared with the parent compound. This correlates with an albeit slower rotational correlation time (τR). Nevertheless, the largest of the complexes, possessing the largest number of methyl and methylene moieties, had the slowest water exchange rate. This compromised any further theoretical enhancement in relaxivity.

8.2 BIODISTRIBUTION

The most important property determining the biodistribution of dendrimer-based MR contrast agents is their size. Of course, the size in turn is determined by the nature of the central core, the internal architecture and the generation number (*G*). The first dendrimer-based MR agents evaluated for their application in MR angiography (MRA) were PAMAM-Am-G2-DTPA and PAMAM-Am-G6-DTPA [3]. These agents had enhancement half-lives that were two- and ten-times more compared with Gd^{3+}-DTPA, respectively, in murine model. A preliminary dose-response experiment

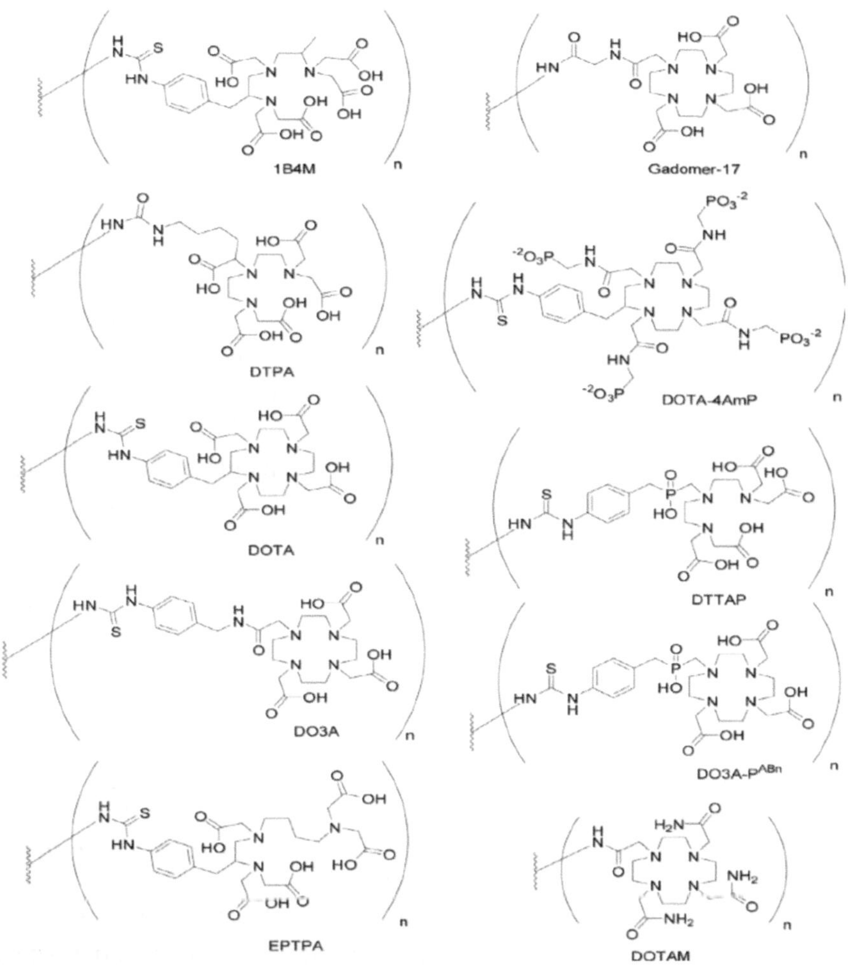

FIGURE 8.1 Chelate-functionalized dendrimers. (From Aaron Joseph L. Villaraza, Ambika Bumb, and Martin W. Brechbiel. Macromolecules, Dendrimers and Nanomaterials in Magnetic Resonance Imaging: The Interplay Between Size, Function and Pharmacokinetics. Chem Rev. 2010 May 12; 110(5): 2921–2959. doi:10.1021/cr900232t.)

in rabbits, described the use of PAMAM-Am-G5-DO3A for visualizing the vasculature, and reported a minimum effective dose and maximal contrast enhancement doses of 0.02 mmol/kg and 0.03 mmol/kg respectively [17]. Further, in the subsequent MRA experiments that involved a series of PAMAM-Am-DO3A (G2–G5), the circulating half-lives of the agents was noted to increase with the increasing generation number [7].

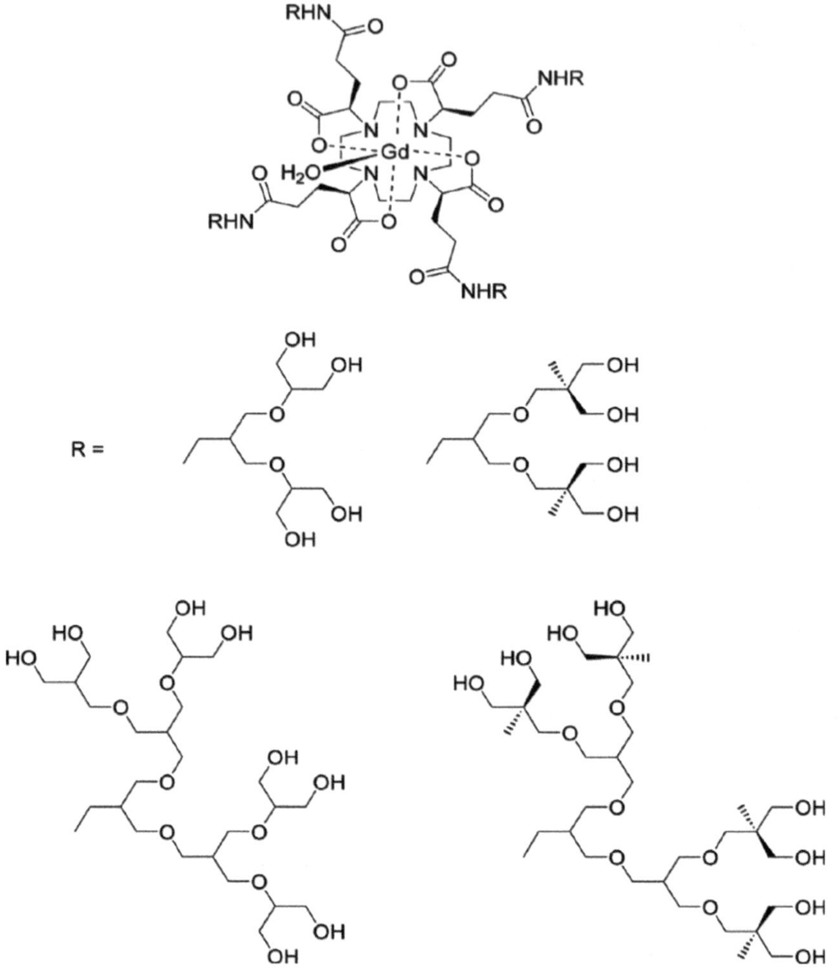

FIGURE 8.2 Gd^{3+} chelate at barycenter of dendrimer. (From Aaron Joseph L. Villaraza, Ambika Bumb, and Martin W. Brechbiel. Macromolecules, Dendrimers and Nanomaterials in Magnetic Resonance Imaging: The Interplay Between Size, Function and Pharmacokinetics. Chem Rev. 2010 May 12; 110(5): 2921–2959. doi:10.1021/cr900232t.)

Gadomer-17 (also known as Gd^{3+}-DTPA-24-cascade polymer; a dendrimer made up of a trimesoyl triamide core and having branched lysine) [18] when comparison with Gd^{3+}-DTPA, poly Lys-DTPA and Gd^{3+}-DTPA-albumin [19–21] was observed to visualize intra-tumoral vasculature and exhibited high vascular permeability [22]. In a murine model, a dose of 0.033 mmol/kg of PAMAM-Am-G6–1B4M was sufficient to visualize intra-tumoral vasculature with a diameter as small as 100 μm [23]. Further, Gadomer-17 (0.1 mmol/kg with 24 Gd^{3+}-DOTA units) has been used to image the vasculature in dogs. The MRA images showed that a more enhanced contrast compared to a 0.3 mmol/kg dose of Gd^{3+}-DTPA [24].

Several researchers undertook systematic studies pertaining to the biodistribution of PAMAM-EDA-1B4M chelating Gd^{3+}-dendrimers, which ranged from G3–G10, for their potential use in MRA studies in order to visualize normal as well as tumoral vasculature [25–28]. Their results revealed that the lower generations (G3–G5) exhibited faster clearance from the body due to a high glomerular filtration rate, even though G5/G6 were retained long enough enabling the researchers to visualize the normal ultrafine vasculature (the limit was up to 200 μm). On the other hand, the higher-generation dendrimers (G7–G9) were observed to have lower renal clearance than G6. Further, G8/G9 exhibited a significantly higher accumulation in the liver. On a temporal basis, G8 could visualize the intra-tumoral blood vessels in a much more stable manner than G6. This was due to the increased vascular permeability known to occur in rapidly-growing tumor cells. In summary, the researchers suggested that G7 was the best candidate as far as visualizing the intra-tumoral vasculature was concerned because it could be retained in the circulation for the longest period. The low hepatic uptake and slow renal clearance have permitted longer time to acquire images. The highest generation dendrimer (G10) precipitated at physiological pH. In a similar way, other workers have studied the biodistribution of lower to intermediate generation PPI-DTPA dendrimers (from core – G4) [29]. All the compounds studied underwent renal clearance, although the higher generations had extended retention in the circulation. Furthermore, G2/G4 demonstrated a lower tendency of leakage from the tumor vasculature into the tumor, whereas the core and G0 did so rapidly. Finally, the largest dendrimer studied (G4) was found to possess a lowest detectable concentration nearly 80 nM. This was more than two orders of magnitude lower compared to that of Gd^{3+}-DTPA. Similar work carried out by other groups documented that between PAMAM-EDA-G6–1B4M and PAMAM-Am-G6–1B4M, the former exhibited longer retention in the blood and slower renal clearance, thereby proving it to be a better blood pool agent [30]. Besides, PPI-G4–1B4M was shown to exhibit significant hepatic uptake when compared with PAMAM-EDA-G4–1B4M which was due to its relatively higher hydrophobic nature [31]. It was also demonstrated to visualize normal liver parenchyma as well as micro-metastatic growths (0.03 mm diameter) in murine model [32]. Owing to the problems of prolonged retention in circulation and reduced clearance, a comparative study between dendrimers of varying sizes and cores was performed to find out which dendrimer had the best renal excretion properties. PAMAM-EDA-G2–1B4M, PPI-G2–1B4M and PPI-G3–1B4M emerged as the best candidates for clinical studies [33]. PPI-G2–1B4M was reported to be the best agent for performing functional renal imaging and early diagnosis of renal disease [34]. Further, control over the circulation as well as the excretion properties could be demonstrated by conjugating the dendrimers with polyethylene glycol (PEG), or co-injection with lysine, or even biotinylation of the dendrimer to be followed by an avidin chase [35–37]. The higher generation dendrimers were observed to be more suitable for applications in MR lymphangiography. Thus, for example, PAMAM-EDA-G8–1B4M (with a large size and hence low vascular permeation) was retained inside the lymphatic compartments, thus permitting to discriminate between infection versus proliferative or neoplastic swelling [27]. Next, a comparison between the dendrimers of varying cores revealed that this same dendrimer-based agent was better suited for imaging the lymphatics while PPI-G5-DTPA was better for visualizing the lymph nodes [38]. Finally, major advances in bioconjugation chemistry have permitted the preparation of multi-modal dendrimer-based fluorescent-MRI probe which is capable of visualizing sentinel lymph nodes [39, 40]. Of late, PAMAM-EDA-G8–1B4M was assessed as a CT-MR probe when co-administered with convection-enhanced delivery (CED) of therapy to the nervous system, even though the effect of dendrimer size/core in this field of use is yet to be determined [41].

8.3 TARGETED AGENTS

Many attempts have been made to enhance the selectivity of dendrimer-based MR agents, mainly by synthesizing targeted bioconjugates. For example, a set of antibody-tagged (mAb 2E4) dendrimers, viz. PAMAM-Am-G2-DOTA and PAMAM-Am-G2-CHXB, were efficiently labeled (using ^{90}Y, ^{111}In, ^{212}Bi, and Gd^{3+}, without compromising the immunoreactivity) as a potential means for targeted radiotherapy or MR imaging [42]. Researchers have demonstrated that by conjugating PAMAM-EDA-G4–1B4M with OST7 (which is a murine monoclonal IgG1), there is no compromise in the immunoreactivity. Further, besides the desired specific accumulation at the tumor sites, the antibody-dendrimer complex had better circulatory clearance than the simple 1B4M-labeled antibody [43]. The PAMAM-Am-G4-DTPA conjugated with folic acid has also been successfully demonstrated to selectively label the ovarian cancer sites that overexpress the high-affinity folate receptor [44–47]. The PAMAM-EDA-G3 has been consecutively conjugated with a fluorescent dye called cyclic-RGD and Gd^{3+}-1B4M for selectively visualizing integrin $\alpha V\beta 3$. The latter is a marker for angiogenesis [48]. Although the *in vitro* results were promising initially, later the approach had limited success *in vivo.*

8.4 CELL TRANSFECTION

Attempts have also been made to develop techniques for the intracellular delivery of various contrast agents. Solution studies of dendrimer-based as well as non-dendrimer-based contrast agents in combination with specific cell transfection agents have revealed that the adduct formation can reduce the relaxivity of Gd^{3+}-based agents. This is due to blocking of the water coordination sites. However, the adduct dissociation was noted to be a function of pH thereby suggesting an added capability of such systems as a pH switch [49]. Successful cellular delivery was reported by workers using a bioconjugate construct constituted of PAMAM-EDA-G6 which was labeled with biotin, Gd^{3+}-1B4M, and finally avidin, and it was found to specifically accumulate in SHIN3 tumor cells (used to study human ovarian cancer). The retention was nearly 50 times greater than the mononuclear Gd^{3+}-DTPA [50]. Other workers employed a 3-step pre-targeting approach in order to visualize Her-2/*neu* xenografts in murine model. In this approach, at first biotinylated trastuzumab was administered for labeling the tumors. This was followed by an avidin chase, and finally biotinylated PAMAM-G4-DTPA dendrimer. Even though a very limited selective MR enhancement could be observed in the tumor xenografts, owing to the

EPR effect the bioconjugate construct could be retained in the tumors [51]. Another research group has described the use of a cysteamine-core dendrimer for producing a multi-modal dendrimer-based agent, by employing intelligent chemistry. PAMAM-CYS-G2 was initially conjugated with 1B4M-DTPA (Gd^{3+}). Subsequently, the disulfide core of the dendrimer was cleaved in order to allow for site-specific biotin conjugation. Four of these bioconjugate constructs could form adducts with avidin (fluorescently-labeled). Multi-modal imaging in mice with ovarian cancer confirmed the accumulation of the supramolecular construct [52].

Convergent synthetic strategies are commonly employed to produce dendrimer-based MR contrast agents. This involves conjugating branched carbohydrates onto a paramagnetic core. Thus, for example, researchers have produced branched aminoglycoside wedges having four and twelve glucose units that were subsequently reacted with DTPA dianhydride. This resulted in a glycodendrimer [53]. Similarly, other workers have reported the synthesis of C-4 symmetric glycoconjugates wherein 4 branched carbohydrate wedges having 12 terminal glucose/galactose units, were linked onto a DOTA core (Figure 8.3) [54]. Like in the case with paramagnetic arborols constructed similarly, the resulting Gd^{3+}-chelate showed a high relaxivity [15]. This was attributed not only to a secondary hydration sphere effect, but also to an enhanced motional coupling owing to the fact that the Gd^{3+} ion resided at the barycenter of the macromolecular complex.

8.5 COMPUTED TOMOGRAPHY AGENTS

Unfortunately, there has been a relatively modest research pertaining to the use of dendrimers as CT contrast agents. This is mainly due to two reasons- (i) the high dose requirement for iodinated contrast agents that is suitable for CT and consequently, the toxicity concern; and (ii) the risk of exposing small laboratory animals to the high dose of ionizing radiation during a CT scan [55]. This is especially problematic in conditions when repeated scans may be necessary, for example in relation to dynamic CT involving multiple scans. It is also an issue in the monitoring of the tumor response following anti-angiogenic treatment. These problems make it difficult to evaluate these formulations in the pre-clinical set up. However, under certain specific circumstances, iodinated CT dendrimer agents

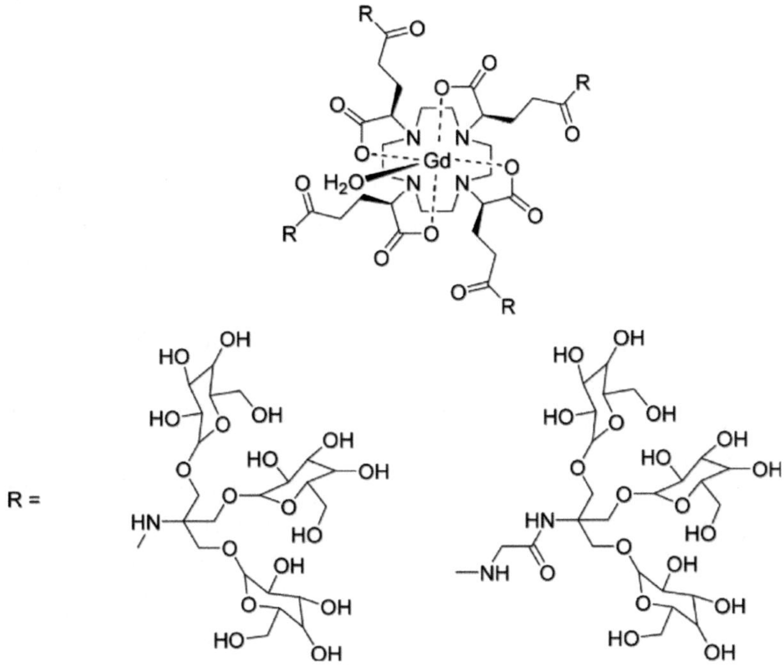

FIGURE 8.3 Glycodendrimers. (From Aaron Joseph L. Villaraza, Ambika Bumb, and Martin W. Brechbiel. Macromolecules, Dendrimers and Nanomaterials in Magnetic Resonance Imaging: The Interplay Between Size, Function and Pharmacokinetics. Chem Rev. 2010 May 12; 110(5): 2921–2959. doi:10.1021/cr900232t.)

may be indeed beneficial. The classical contrast-enhanced CT scans are based on using the low-molecular weight iodinated agents (typically <2 KDa). These agents can readily leak from the highly permeable tumor vasculature. However, they do also have high first-pass extravasations via the normal non-cranial/non-cerebral blood vessels. CT angiography is employed to visualize the arterial and venous blood flow. Important indications include pulmonary angiography (in case of a suspected pulmonary thrombus/embolus) and neuro-angiography (in case of a suspected cerebrovascular accident). Contrast agents having a large molecular weight have the added benefit of being confined within the intravascular compartment and are therefore more suited as blood pool enhancing agents in comparison to their low molecular weight counterparts. With this aim in mind, macromolecular contrast agents have been started to be developed. P743 is a non-ionic iodinated macromolecule (molecular weight

13 kD) which was shown to possess pharmacokinetic as well as imaging profiles that were remarkably consistent with those expected from a rapid-clearance blood-pool agent [56]. Attempts to develop larger macromolecular agents (>30 kDa) for the purpose of CT enhancement, that could achieve a higher blood pool retention are still short of reaching the stage of clinical trials. Such examples include the polymeric micelles synthesized from iodine-containing amphiphilic block-copolymers, as well as the iopromide-containing liposomes (PEGylated or non-PEGylated) [57, 58]. However, both these complexes display a high polydispersity. This is likely to compromise both reproducibility and accuracy. Therefore, preferably low polydispersity dendrimer-based compounds would be more attractive as macromolecular CT contrast agents. Like the other macromolecules developed for CT imaging, the dendrimers are obviously not inherently radio-opaque. Therefore, they need to be coupled to iodinated compounds at the surface. X-ray CT is significantly less sensitive than MRI, hence high concentrations of iodine are required. This makes the chemical synthesis of macromolecular contrast agents an uphill task. As observed for Gadolinium-dendrimers, the extended retention of iodinated dendrimeric agents may also be problematic. The major adverse effects of iodinated agents include anaphylactoid reactions, and renal damage even causing acute renal failure [59, 60]. For these reasons, iodinated agents are contra-indicated in subjects with pre-existing renal compromise, or in subjects who are at significant risk of such damage. The patients undergoing dialysis are exempt, since dialysis can be used to remove the agent [61].

Researchers have recently developed a paired dendrimer particle which is CT compatible and is based on a PEG core [62]. Paired G3, G4 and G5 dendrimers were synthesized with molecular weight ranging from 35 – 143 kDa, and having 16, 32, and 64 amino groups, respectively, available to conjugate with the reactive tri-iodophthalamide units. In rat models, the dendrimers were able to attain successful imaging with intravascular enhancement. The half-life was 35 minutes for G4 molecules. All the complexes evaluated demonstrated high water-solubility/hydrophilicity, low osmolality, and excellent chemical stability. Although CT is known to have the distinct demerit of radiation exposure, the macromolecular agents may eventually prove worthy for certain types of CT investigations.

8.6 OPTICAL IMAGING AGENTS

Optical imaging has several advantages compared to the other imaging modalities (Table 8.1) [55, 63].

However, they have limited ability to penetrate through tissue and this is a major problem for *in vivo* human studies. Plenty of research has focused on using dendrimers in optical imaging, although most of these studies have used *in vitro* models. It has been observed before that PAMAM dendrimers having carboxyl and amino terminals have a weak though detectable intrinsic fluorescence [64]. In addition, the simple oxidation of OH-terminated PAMAM dendrimers can produce high quantum yield blue fluorescence [65]. Researchers have observed a strong fluorescence emission from a variety of dendrimers at acidic pH [66]. Dendrimer-stabilized gold particles were shown to display strong blue emission at 458 nm [67]. This inherent fluorescence was exploited, using G6 dendrimers, to monitor their uptake and movement within the cytosol of target cells, without the necessity for linking to any external optical contrast agents [68]. The inherent fluorescence of dendrimers has its distinct advantages (Table 8.2).

TABLE 8.1 Advantages of Optical Imaging Compared to Other Imaging Modalities

- Portable cameras.
- Relatively inexpensive.
- High sensitivity and specificity.
- Excellent temporal and spatial resolution.
- No exposure to ionizing radiation.

TABLE 8.2 Advantages of Inherent Fluorescence of Dendrimers in Optical Imaging

- No need to incorporate potentially injurious optical agents, e.g., inorganic, cadmium-based dyes.
- Size of the unmodified dendrimer is not affected.
- Mobility of the unmodified dendrimer is unaltered.
- No risk of the attached optical probe dissociating before the dendrimer reaches the target.

Reports have shown the binding/uptake of FITC (fluorescein isothiocyanate)-conjugated PAMAM G3 dendrimers into cells, using flow-cytometry as well as confocal laser scanning microscopy [69]. The TEM (transmission electron microscopy) analysis of cells after incubing with gold-tagged PAMAM G3 dendrimers did confirm the endocytosis-mediated internalization. It has been demonstrated that G5 dendrimer-entrapped gold nanoparticles could be covalently attached to folic acid and FITC for targeting and optical imaging of cancer cells [70]. Dendrimers were used to amplify the DNA microarray fluorescent signals for detecting viral pathogens [71]. G4-PAMAM dendrimers have been conjugated with Cy3-flourophores and the pertinent complementary oligonucleotide probes so as to generate sufficiently amplified fluorescence intensity while detecting herpes simplex virus. Researchers have also successfully used FITC-labeled monoclonal antibody conjugated PAMAM G5 dendrimers to assess the targeting/uptake of through the cell surface receptors CD14 as well as the prostate-specific antigen *in vitro* [72]. An 'Alexa' Fluor 594-labeled G4-PAMAM has been developed for the almost continuous detection of blood glucose levels in diabetic patients. It is an implantable chemical assay with a potential clinical application. [73]. Rather similar dextran-based assays, also exploiting optical imaging, operate through a competitive binding reaction between concanavalin A, dextran, and glucose. The dendrimer-based assay was reported to be advantageous with respect to having a greater response to physiological blood glucose concentrations, besides incorporating longer wavelength dyes for improving signal penetration. PAMAM G5 dendrimers conjugated to FITC as well as peptides containing Arg-Gly-Asp (RGD) motif targeting the integrin receptors $\alpha v/\beta 3$ [74] have also been developed. The $\alpha v/\beta 3$ receptors are known to be preferentially expressed on endothelial cells proliferating during angiogenesis [75]. These dendrimer-based platforms with specificity for $\alpha v/\beta 3$ integrins would be useful for the *in vivo* evaluation of tumor angiogenesis and monitoring the response to anti-angiogenic drugs [76]. A PAMAM G5 dendrimer labeled with 'Alexa' Fluor 488, conjugated with anti-HER2 mAb (monoclonal antibody) was evaluated [77]. Overexpression of HER2 receptors is known to occur in many epithelial tumors such as breast cancer, and is often associated with an aggressive tumor behavior [78–80]. High expression of HER2 serves to identify patients who would most likely benefit from treatment with trastuzumab, which is an anti-HER2 monoclonal antibody [81]. Studies have shown successful targeting of the dendrimer-mAb conjugate to the HER2 receptors besides a rapid yet more efficient internalization when compared with anti-HER2 antibody alone [77]. The potential future applications of these complexes include the optimization of targeted drug delivery against HER2 positive tumors. Optically-labeled G4 dendrimers were also developed for industrial use, specifically as sensors for detecting microbial contamination in water supplies [82]. However, till date, relatively few groups have actually succeeded in adapting optical dendrimers for use in the *in vivo* models.

Researchers have developed a PAMAM G4 dendrimer core covalently linked to a specific fluorescein-labeled, polymer-based substrate called PB-M7VIS. The latter targets matrix metalloproteinase-7 (MMP-) [83]. The MMP-7 is known to be a key proteinase in the spread of a number of tumors (such as colon and breast) because it is involved in the remodeling of neoplastic blood vessels. Hence, it may be a suitable target for the therapy of such tumors. After it binds, the MMP-7 cleaves the peptide resulting in a several-fold increment in the optical signal of fluorescein. A murine xenograft model of human colon cancer cells were selected, either positive/negative for MMP-7. It was observed that the PBM7VIS-dendrimers generated a significantly higher fluorescence signal in MMP-7-positive tumors, in contrast to the controls, i.e., MMP-7 negative tumors. In addition, prior treatment of the MMP-7 positive tumors using an MMP inhibitor significantly reduced the fluorescence of MMP-7 positive tumors thereby documenting the selective nature of the optical dendrimer. In other *in vivo* models (cremaster muscles), the process of extravasation of the fluorescently tagged (FITC) PAMAM dendrimers (G0–G4) was studied [84]. The muscles were visualized for leakage using intravital microscopy. It was observed that with an increase in dendrimer size, there was an exponential increase in the time required for the extravasation to occur. Optical dendrimers have also been used to monitor the dendrimeric delivery of genes to the nucleus of cells as well as in dual-modality imaging.

It is clear that though optical imaging has its own distinct merits, yet major hurdles need to be overcome before it may be successfully translated into clinical practice.

8.7 NUCLEAR MEDICINE AGENTS

Nuclear medicine deals with radionuclides or radio-pharmaceuticals (i.e., pharmaceuticals labeled with radionuclides) that are administered to patients and the emitted radiation is detected by suitable sensors. A gamma camera is a good example of a sensor/detector. In scintigraphic imaging, the camera detects the emission of the radionuclide in a planar fashion, whereas in SPECT (Single Photon Emission Tomography), the camera is capable of detecting and reconstructing the emissions in the form of 3-dimensional images [85]. On the other hand, PET (Positron Emission Tomography) uses compounds that have been labeled with positron-emitting radionuclides in order to provide images of specific physiologic processes. It is based on the detection of 2 photons produced when a positron emits from the nucleus of an unstable radionuclide followed by annihilation with its antiparticle (i.e., an electron). Remarkably, PET has a 10-fold higher sensitivity compared with SPECT and it can detect a tracer in picomolar concentrations. However, both SPECT as well as PET are limited by a low spatial resolution. Interestingly, PET or SPECT may be co-acquired with CT scans during the same session so as to provide image attenuation correction, besides allowing a more accurate correlation of radionuclide uptake [86–88]. Depending on the efficiency of the targeting process, it has been observed that low signal-to-noise ratios make it tough to distinguish the targets from the background. Due to this, the majority of research related to the use of dendrimers in the area of nuclear medicine has concentrated albeit less on the primary diagnosis of tumors, but more on the treatment, e.g., dendrimer-based systems delivering cytotoxic radiation (a type of radio-immunotherapy) [55].

8.8 TRANSMISSION ELECTRON MICROSCOPY AGENTS

The development of DENP (dendrimer- entrapped metal nanoparticles) has been a field of interest. High electron density properties of noble metals (e.g., gold, silver) allows the imaging with TEM (transmission electron microscopy). On the other hand, the interactions with surrounding microenvironment are actually determined by the attributes of the hosting dendrimer. Researchers have studied gold/PAMAM dendrimers both *in vitro* and *in vivo* and demonstrated that a few atoms of gold for every PAMAM dendrimer already enable the visualization of amorphous single/multiple nano-composite units in biological systems by using TEM [89]. The synthesis of silver/dendrimer nano-composites has been reported with the potential for TEM imaging [90, 91]. Following the incubation of poly-cationic/polyanionic Ag/dendrimers with various cell lines, the particles could be detected on the biomembrane surfaces, in the cytosol, or even trapped inside the phagocytic/endocytic vesicles. However, the cellular uptake of the electrically neutral composites was low. TEM has also been utilized for demonstrating the *in vitro* binding/internalization of Au-DENP (dendrimer-entrapped gold nanoparticles) attached to folic acid and FITC in cells overexpressing folate receptors [70].

8.9 MULTI-MODALITY AGENTS

Every imaging modality has its own distinct merits and limitations. The concurrent use of two or more modalities might help to overcome the demerits of the individual techniques, thereby adding and refining the information gathered during a single session. As already mentioned, the combined usage of PET and CT is a good example of multi-modal imaging. The combination of the two modalities can assist in identifying and localizing the functional abnormalities [92]. Generally speaking, the imaging agents can be detected only by a specific modality and the concurrent usage of two separate agents is rarely necessary. Thus, for instance, PET-CT scans may be acquired with a PET agent alone and depending on the clinical query being addressed, the unenhanced CT images might be suboptimal.

Dendrimers, having multiple binding sites and with the potential to take up multiple, separate imaging agents as the payload, enable their detection by 2 or more modalities. Such probes can further enhance the information gathered from multi-modality scans. Till date there exists only selected instances of dual-modality probes that were successfully developed (Table 8.3) [93–96].

Fluorescence studies have shown that conjugation of Gd ions did not alter the quantum yield of the probe. However, a rise in the Cy5.5 dye content resulted in the partial quenching of the fluorophor. This dendrimer-based novel dual modality probe was successfully tested to visualize the

TABLE 8.3 Dual-Modality Probes

- Peptide-based probes for recognition by radionuclide as well as optical imaging.
- Probes based on the macromolecular agent, albumin-GdDTPA.
- Linking fluorescent probes for combined MR/optical imaging monitoring of lymph node metastases.
- G6-PAMAM dendrimer nanoprobe complexed to Gd ions/Cy5.5 to allow recognition by MR and near infrared fluorescence imaging.

SLNs (sentinel lymph nodes) in murine models. Advancing this work, other researchers used a similar probe to identify as well as resect the SLNs during near infrared, optical image-guided surgery [97]. Owing to the high background noise, the near infrared optical imaging failed to identify some SLNs situated near the injection site of the agent. However, MR lymphangiography was able to overcome this issue and permitted accurate localization of all the SLNs irrespective of the location. This kind of a probe may prove beneficial for the peri-operative detection and/or removal of SLNs in clinical scenarios like primary breast cancer, malignant melanoma, etc.

Another good combination of dendrimer-based dual modality imaging agent is a radioisotope/5-color near infrared fluorophor-labeled PAMAM G6 dendrimer that has been prepared and used for lymphatic imaging [98]. Nearly similar sensitivity of radionuclide and fluorescence permitted to significantly minimize the dose of the injected agent in comparison to the above mentioned MR/optical agent used to visualize the lymphatic system. Thus, multi-modality contrast agents have distinct advantages and macromolecular nano-probes especially dendrimers are ideally suited to incorporate multiple imaging beacons per carrier molecule.

8.10 NOW AND NEXT?

The use of macromolecular MR agents for studying and modulating various biological events is certainly an exciting and dynamic domain of research for both basic knowledge and clinical applications. It is remarkable that more than one thousand research/review articles on dendrimers alone have been published in reputed journals in the last 4 years. A critical analysis of this information reveals certain common themes which enable dendrimers

TABLE 8.4 Common Themes Favoring the Diagnostic Use of Dendrimers

• Architectural control.
• Well-defined structure.
• Modifiable functionality.
• Tailored bio-elimination.
• Consistent clinical grade synthesis.

TABLE 8.5 Other Useful Applications of Dendrimer-Based Diagnostic Tools

• Proteomics.
• Immunodiagnostics.
• Pathogen pacification.
• Gene transfection.
• Drug delivery.

in gaining researchers' confidence and faith for diagnostic applications (Table 8.4). Although many of the evaluated dendrimeric complexes have proven their worth in diagnostics, in particular as MR-contrast agents, their future lies in their ability to combine with other agents for improving the diagnostics and, even for therapeutics [91–98]. This platform technology also opens doors for specific and customized combinational applications in several other fields of clinical relevance (Table 8.5). It is reasonable to expect that dendrimer-based diagnostics shall play a key role in the development of novel biomedical devices, contrast agents, and evolving further strategies for the better management of human disease.

KEYWORDS

- **biodistribution**
- **multi-modality agents**
- **nuclear medicine**
- **optical imaging**
- **tomography**
- **transfection**

REFERENCES

1. Andrä, W., & Nowak, H. (2006). *Magnetism in Medicine: A Handbook.* Wiley-VCH; Berlin.
2. Bernard, V. (1989). Australas. *Radiol 33*, 390.
3. Wiener, E., Brechbiel, M. W., Brothers, H., Magin, R. L., Gansow, O. A., Tomalia, D. A., & Lauterbur, P. C. (1994). *Magn. Reson. Med 31*, 1.
4. Tomalia, D. A., Baker, H., Dewald, J., Hall, M., Kallos, G., Martin, S., Roeck, J., Ryder, J., & Smith, P. (1985). *Polym. J 17*, 117.
5. Xu, H., Regino, C. A. S., Bernardo, M., Koyama, Y., Kobayashi, H., Choyke, P. L., & Brechbiel, M. W. (2007). *J. Med.Chem 50*, 3185.
6. Langereis, S., de Lussanet, Q. G., van Genderen, M. H. P., Backes, W. H., & Meijer, E. W. (2004). *Macromolecules 37*, 3084.
7. Margerum, L. D., Campion, B. K., Koo, M., Shargill, N., Lai J-J, Marumoto, A., & Sontum, P. C. (1997). *J. Alloys Compd 249*, 185.
8. Bryant, L. H. J., Brechbiel, M. W., Wu, C., Bulte, J. W. M., Herynek, V., & Frank, J. A. (1999). *J. Magn. Reson. Imaging 9*, 348.
9. Tóth, E., Pubanz, D., Vauthey, S., Helm, L., & Merbach, A. E. (1996). *Chem. Eur. J 2*, 1607.
10. Laus, S., Sour, A., Ruloff, R., Tóth, E., & Merbach, A. E. (2005). *Chem. Eur. J 11*, 3064.
11. Laus, S., Ruloff, R., Tóth, E., & Merbach, A. E. (2003). *Chem. Eur. J 9*, 3555.
12. Lebduskova, P., Sour, A., Helm, L., Toth, E., Kotek, J., Lukes, I., & Merbach, A. E. (2006). *Dalton Trans 28,* 3399.
13. Pikkemaat, J. A., Wegh, R. T., Lamerichs, R., van de Molengraaf, R. A., Langereis, S., Burdinski, D., Raymond, A. Y. F., Janssen, H. M., de Waal, B. F. M., Willard, N. P., Meijer, E. W., & Grüll, H. (2007). *Contrast Media Mol. Imaging 2*, 229.
14. Aime, S., Carrera, C., Delli Castelli, D., Crich, S. G., & Terreno, E. (2005). *Angew. Chem., Int. Ed 44*, 1813.
15. Fulton, D. A., O'Halloran, M., Parker, D., Senanayake, K., Botta, M., & Aime, S. (2005). *Chem. Commun 4, 474.*
16. Merbach, A. E., & Toth, E., editors. The Chemistry of Contrast Agents in Medical Magnetic Resonance Imaging. Wiley; Chichester: 2001.
17. Bourne, M. W., Margerun, L., Hylton, N., Campion, B., Lai J-J, Derugin, N., & Higgins, C. B. (1996). *J. Magn. Reson. Imaging 6*, 305.
18. Weinmann, H. J., Ebert, W., Wagner, S., Taupitz, M., Misselwitz, M., & Schmitt-Wilich, H. (1997). Proceedings of the IX International Workshop on Magnetic Resonance Angiography Valencia; p. 355.
19. Adam, G., Neuerburg, J., Spüntrup, E., Mühler, A., Scherer, K., & Günther, R. W. (1994). *J. Magn. Reson. Imaging 4*, 462.
20. Roberts, H. C., Saeed, M., Roberts, T. P. L., Mühler, A., Shames, D. M., Mann, J. S., Stiskal, M., F. D., & Brasch, R. C. (1997). *J. Magn. Reson. Imaging 7,* 331.
21. Tacke, J., Adam, G., Claßen, H., Mühler, A., Prescher, A., & Günther, R. W. J. Magn. Reson. Imaging (1997). 7, 678.

22. Su, M-Y, Mühler, A., Lao, X., & Nalcioglu, O. (1998). *Magn. Reson. Imaging 39*, 259.
23. Kobayashi, H., Sato, N., Kawamoto, S., Saga, T., Hiraga, A., Ishimori, T., Konishi, J., Togashi, K., & Brechbiel, M. W. (2001). *Magn. Reson. Med 46,* 579.
24. Dong, Q., Hurst, D. R., Weinmann, H. J., Chevenert, T. L., Londy, F. J., & Prince, M. R. (1998). *Invest. Radiol 33*, 699.
25. Sato, N., Kobayashi, H., Hiraga, A., Saga, T., Togashi, K., Konishi, J., & Brechbiel, M. W. (2001). *Magn. Reson. Med 46,* 1169.
26. Kobayashi, H., Kawamoto, S., Star, R. A., Waldmann, T. A., Brechbiel, M. W., & Choyke, P. L. (2003). *Bioconjugate Chem 14*, 1044.
27. Kobayashi, H., Kawamoto, S., Star, R. A., Waldmann, T. A., Tagaya, Y., & Brechbiel, M. W. (2003). *Cancer Res 63*, 271.
28. Yordanov, A., Kobayashi, H., English, S. J., Reijnders, K., Milenic, D., Krishna, M. C., Mitchell, J. B., & Brechbiel, M. W. (2003). *J. Mater. Chem 13*, 1523.
29. Langereis, S., de Lussanet, Q. G., van Genderen, M. H. P., Meijer, E. W., Beets-Tan, R. G. H., Griffioen, A. W., van Engelshoven, J. M. A., & Backes, W. H. (2006). *NMR Biomed 19*, 133.
30. Kobayashi, H., Sato, N., Kawamoto, S., Saga, T., Hiraga, A., Laz Haque, T., Ishimori, T., Konishi, J., Togashi, K., & Brechbiel, M. W. (2001). *Bioconjugate Chem 12*, 100.
31. Kobayashi, H., Kawamoto, S., Saga, T., Sato, N., Hiraga, A., Ishimori, T., Akita, Y., Mamede, M. H., Konishi, J., Togashi, K., & Brechbiel, M. W. (2001). *Magn. Reson. Med 46*, 795.
32. Kobayashi, H., Saga, T., Kawamoto, S., Sato, N., Hiraga, A., Ishimori, T., Konishi, J., Togashi, K., & Brechbiel, M. W. (2001). *Cancer Res 61*, 4966.
33. Kobayashi, H., Kawamoto, S., Jo, S., Bryant, L. H., Brechbiel, M. W., & Star, R. A. (2003). *Bioconjugate Chem 14*, 388.
34. Kobayashi, H., Kawamoto, S., Saga, T., Sato, N., Hiraga, A., Ishimori, T., Konishi, J., & Togashi, K., W. (2001). BM. *Magn. Reson. Med 46,* 781.
35. Kobayashi, H., Sato, N., Kawamoto, S., Saga, T., Hiraga, A., Ishimori, T., Konishi, J., & Togashi, K., Brechbiel, M. W. (2001). *Magn. Reson. Med 46*, 457.
36. Kobayashi, H., Kawamoto, S., Star, R. A., Waldmann, T. A., Brechbiel, M. W., & Choyke, P. L. (2003). *Bioconjugate Chem 14*, 1044.
37. Kobayashi, H., Kawamoto, S., Choyke, P. L., Sato, N., Knopp, M. V., Star, R. A., Waldmann, T. A., Tagaya, Y., & Brechbiel, M. W. (2003). *Magn. Reson. Med 50*, 758.
38. Koyama, Y., Talanov, V. S., Bernardo, M., Hama, Y., Regino, C. A. S., Brechbiel, M. W., Choyke, P. L., & Kobayashi, H. J. (2007). *Magn. Reson. Imaging 25*, 866.
39. Talanov, V. S., Regino, C. A. S., Kobayashi, H., Bernardo, M., Choyke, P. L., & Brechbiel, M. W. (2006). *Nano Lett 6,* 1459.
40. Regino, C. A. S., Walbridge, S., Bernardo, M., Wong, K. J., Johnson, D., Lonser, R., Oldfield, E. H., Choyke, P. L., & Brechbiel, M. W. (2008). *Contrast Media Mol. Imaging 3*, 2.
41. Wu, C., Brechbiel, M. W., Kozak, R. W., & Gansow, O. A. (1994). *Bioorg. Med. Chem 4*, 449.
42. Kobayashi, H., Sato, N., Saga, T., Nakamoto, Y., Ishimori, T., Toyama, S., Tagashi, K., Konishi, J., & Brechbiel, M. W. (2000). *Eur. J. Nucl. Med 27*, 1334.

43. Konda, S., Aref, M., Brechbiel, M. W., & Wiener, E. C. (2000). *Invest. Radiol 35*, 50.
44. Konda, S. D., Aref, A., Wang, S., Brechbiel, M. W., & Wiener, E. C. (2001). *MAGMA 12,* 104.
45. Konda, S. D., Wang, S., Brechbiel, M. W., & Weiner, E. C. (2002). *Invest. Radiol 37,* 199.
46. Wiener, E. C., Konda, S., Shadron, A., Brechbiel, M., & Gansow, O. (1997). *Invest. Radiol 32*, 748.
47. Boswell, C. A., Eck, P. K., Regino, C. A. S., Bernardo, M., Wong, K. J., Milenic, D. E., Choyke, P. L., & Brechbiel, M. W. (2008). *Mol. Pharm 5*, 527.
48. Kalish, H., Arbab, A. S., Miller, B. R., Lewis, B. K., Zywicke, H. A., Bulte, J. W. M., Bryant, L. H. Jr, & Frank, J. A. (2003). *Magn. Reson. Med 50*, 275.
49. Kobayashi, H., Kawamoto, S., Saga, T., Sato, N., Ishimori, T., Konishi, J., Ono, K., Togashi, K., & Brechbiel, M. W. (2001). *Bioconjugate Chem 12*, 587.
50. Zhu, W., Okollie, B., Bhujwalla, Z. M., & Artemov, D. (2008). *Magn. Reson. Med 59*, 679.
51. Xu, H., Regino, C. A. S., Koyama, Y., Hama, Y., Gunn, A. J., Bernardo, M., Kobayashi, H., Choyke, P. L., & Brechbiel, M. W. (2007). *Bioconjugate Chem 18,* 1474.
52. Takahashi, M., Hara, Y., Aoshima, K., Kurihara, H., Oshikawa, T., & Yamashita, M. (2000). *Tetrahedron Lett 41*, 8485.
53. Fulton, D. A., Elemento, E. M., Aime, S., Chaabane, L., Botta, M., & Parker, D. (2006). *Chem. Commun 10*, 1064.
54. Barrett, T., Ravizzini, G., Choyke, P. L., & Kobayashi, H. (2009). Dendrimers Application Related to Bioimaging. *IEEE Eng Med Biol Mag. 28*(1), 12–22.
55. Idee, J. M., Port, M., Robert, P., Raynal, I., Prigent, P., Dencausse, A., Le Greneur, S., Tichkowsky, I., Le Lem, G., Bourrinet, P., Mugel, T., Benderbous, S., Devoldere, L., Bourbouze, R., Meyer, D., Bonnemain, B., & Corot, C. (2001). Preclinical profile of the monodisperse iodinated macromolecular blood pool agent P743. *Invest Radiol 36*, 41–49.
56. Trubetskoy, V. S., Gazelle, G. S., Wolf, G. L., & Torchilin, V. P. (1997). Block-copolymer of polyethylene glycol and polylysine as a carrier of organic iodine: design of long-circulating particulate contrast medium for X-ray computed tomography. *J Drug Target 4*, 381–388.
57. Sachse, A., Leike, J. U., Schneider, T., Wagner, S. E., Rossling, G. L., Krause, W., & Brandl, M. (1997). Biodistribution and computed tomography blood-pool imaging properties of polyethylene glycol-coated iopromide-carrying liposomes. *Invest Radiol 32*, 44–50.
58. Schild, H. H., Kuhl, C. K., Hubner-Steiner, U., Bohm, I., & Speck, U. (2006). Adverse Events after Unenhanced and Monomeric and Dimeric Contrast-enhanced CT: A Prospective Randomized Controlled Trial. *Radiology 240*, 56–64.
59. Finn, W. F. (2006). The clinical and renal consequences of contrast-induced nephropathy. *Nephrol Dial Transplant 21,* 2–10.
60. Toprak, O. (2007). Conflicting and New Risk Factors for Contrast Induced Nephropathy. *Journal of Urology 178*, 2277–2283.
61. Fu, Y., Nitecki, D. E., Maltby, D., Simon, G. H., Berejnoi, K., Raatschen, H. J., Yeh, B. M., Shames, D. M., & Brasch, R. C. (2006). Dendritic iodinated contrast agents

with PEG-cores for CT imaging: synthesis and preliminary characterization. *Bioconjug Chem 17*, 1043–1056.
62. Graves, E. E., Weissleder, R., & Ntziachristos, V. (2004). Fluorescence molecular imaging of small animal tumor models. *Curr Mol Med 4,* 419–430.
63. Larson, C. L., & Tcker, S. A. (2001). Intrinsic fluorescence of carboxylateterminated polyamidoamine dendrimers. *Applies Spectroscopy, 55*.
64. Lee, W. I., Bae, Y., & Bard, A. J. (2004). Strong blue photoluminescence and ECL from OH-terminated PAMAM dendrimers in the absence of gold nanoparticles. *J Am Chem Soc 126,* 8358–8359.
65. Wang, D., & Imae, T. (2004). Fluorescence emission from dendrimers and its pH dependence. *J Am Chem Soc 126*, 13204–13205.
66. Shi, X., Ganser, T. R., Sun, K., Balogh, L. P., & Baker, J. R. Jr. (2006). Characterization of crystalline dendrimerstabilized gold nanoparticles. *Nanotechnology 17,* 1072–1078.
67. Al-Jamal, K. T., Ruenraroengsak, P., Hartell, N., & Florence, A. T. (2006). An intrinsically fluorescent dendrimer as a nanoprobe of cell transport. *J Drug Target 14*, 405–412.
68. Jevprasesphant, R., Penny, J., Attwood, D., & D'Emanuele, A. (2004). Transport of dendrimer nanocarriers through epithelial cells via the transcellular route. *J Control Release 97*, 259–267.
69. Shi, X., Wang, S., Meshinchi, S., Van Antwerp, M. E., Bi, X., Lee, I., & Baker, J. R. Jr. (2007). Dendrimer-entrapped gold nanoparticles as a platform for cancer-cell targeting and imaging. *Small 3*, 1245–1252.
70. Striebel, H. M., Birch-Hirschfeld, E., Egerer, R., Foldes-Papp, Z., Tilz, G. P., & Stelzner, A. (2004). Enhancing sensitivity of human herpes virus diagnosis with DNA microarrays using dendrimers. *Exp Mol Pathol 77,* 89–97.
71. Thomas, T. P., Patri, A. K., Myc, A., Myaing, M. T., Ye, J. Y., Norris, T. B., & Baker, J. R. Jr. (2004). In vitro targeting of synthesized antibody-conjugated dendrimer nanoparticles. *Biomacromolecules 5*, 2269–2274.
72. Ibey, B. L., Beier, H. T., Rounds, R. M., Cote, G. L., Yadavalli, V. K., & Pishko, M. V. (2005). Competitive binding assay for glucose based on glycodendrimer-fluorophore conjugates. *Anal Chem 77,* 7039–7046.
73. Hill, E., Shukla, R., Park, S. S., & Baker, J. R. Jr. (2007). Synthetic PAMAM-RGD conjugates target and bind to odontoblast-like MDPC 23 cells and the predentin in tooth organ cultures. *Bioconjug Chem 18,* 1756–1762.
74. Cox, D., Aoki, T., Seki, J., Motoyama, Y., & Yoshida, K. (1994). The Pharmacology of the Integrins. *Medicinal Research Reviews 14*, 195–228.
75. Hsu, A. R., Veeravagu, A., Cai, W., Hou, L. C., Tse, V., & Chen, X. Y. (2007). Integrin alpha(v)beta(3) antagonists for anti-angiogenic cancer treatment. *Recent Patents on Anti-Cancer Drug Discovery 2,* 143–158.
76. Shukla, R., Thomas, T. P., Peters, J. L., Desai, A. M., Kukowska-Latallo, J., Patri, A. K., Kotlyar, A., & Baker, J. R. Jr. (2006). HER2 specific tumor targeting with dendrimer conjugated anti-HER2 mAb. *Bioconjug Chem 17,* 1109–1115.
77. Koeppen, H. K., Wright, B. D., Burt, A. D., Quirke, P., McNicol, A. M., Dybdal, N. O., Sliwkowski, M. X., & Hillan, K. J. (2001). Overexpression of HER2/neu in solid tumors: an immunohistochemical survey. *Histopathology 38*, 96–104.

78. Owens, M. A., Horten, B. C., & Da Silva, M. M. (2004). HER2 amplification ratios by fluorescence in situ hybridization and correlation with immunohistochemistry in a cohort of 6556 breast cancer tissues. *Clin Breast Cancer 5*, 63–69.
79. Slamon, D. J., Clark, G. M., Wong, S. G., Levin, W. J., Ullrich, A., & McGuire, W. L. (1987). Human breast cancer: correlation of relapse and survival with amplification of the HER-2/neu oncogene. *Science 235,* 177–182.
80. Hudis, C. A. (2007). Trastuzumab–mechanism of action and use in clinical practice. *N Engl J Med 357,* 39–51.
81. Ji, J., Schanzle, J. A., & Tabacco, M. B. (2004). Real-time detection of bacterial contamination in dynamic aqueous environments using optical sensors. *Anal Chem 76,* 1411–1418.
82. McIntyre, J. O., Fingleton, B., Wells, K. S., Piston, D. W., Lynch, C. C., Gautam, S., & Matrisian, L. M. (2004). Development of a novel fluorogenic proteolytic beacon for in vivo detection and imaging of tumor-associated matrix metalloproteinase-7 activity. *Biochem J 377*, 617–628.
83. El-Sayed, M., Kiani, M. F., Naimark, M. D., Hikal, A. H., & Ghandehari, H. (2001). Extravasation of poly (amidoamine) (PAMAM) dendrimers across microvascular network endothelium. *Pharm Res 18*, 23–28.
84. Miller, T. R. (1996). The AAPM/RSNA physics tutorial for residents. Clinical aspects of emission tomography. *Radiographics 16*, 661–668.
85. Blankespoor, S. C., Wu, X., Kalki, K., Brown, J. K., Tang, H. R., Cann, C. E., & Hasegawa, B. H. (1996). Attenuation correction of spect using x-ray CT on an emission transmission CT system: myocardial perfusion assessment. *IEEE Transactions on Nuclear Science 43,* 2263–2274.
86. Townsend, D. W., Beyer, T., & Blodgett, T. M. (2003). PET/CT scanners: A hardware approach to image fusion. *Seminars in Nuclear Medicine 33,* 193–204.
87. Yeung, H., Schöder, H., Smith, A., Gonen, M., & Larson, S. (2005). Clinical Value of Combined Positron Emission Tomography/Computed Tomography Imaging in the Interpretation of 2-Deoxy-2-[F-18]fluoro-d -glucose–Positron Emission Tomography Studies in Cancer Patients. *Molecular Imaging and Biology 7,* 229–235.
88. Bielinska, A., Eichman, J. D., Lee, I., Baker, J. R. Jr., & Balogh, L. (2002). Imaging Gold-dendrimer Nanocomposites in Cells. *Journal of Nanoparticle Research, 4.*
89. Lesniak, W., Bielinska, A. U., Sun, K., Janczak, K. W., Shi, X., Baker, J. R. Jr., & Balogh, L. P. (2005). Silver/dendrimer nanocomposites as biomarkers: fabrication, characterization, in vitro toxicity, and intracellular detection. *Nano Lett 5,* 2123–2130.
90. Villaraza, A. J. L., Bumb, A., & Brechbiel, M. W. (2010). Macromolecules, Dendrimers and Nanomaterials in Magnetic Resonance Imaging: The Interplay between Size, Function and Pharmacokinetics. *Chem Rev. 110*(5), 2921–2959.
91. Blodgett, T. M., Meltzer, C. C., & Townsend, D. W. (2007). PET/CT: form and function. *Radiology 242*, 360–85.
92. Bullok, K. E., Dyszlewski, M., Prior, J. L., Pica, C. M., Sharma, V., & Piwnica-Worms, D. (2002). Characterization of novel histidine-tagged Tat-peptide complexes dual-labeled with (99m)Tc-tricarbonyl and fluorescein for scintigraphy and fluorescence microscopy. *Bioconjug Chem 13,* 1226–37.

93. Houston, J. P., Ke, S., Wang, W., Li, C., & Sevick-Muraca, E. M. (2005). Quality analysis of in vivo near-infrared fluorescence and conventional gamma images acquired using a dual-labeled tumor-targeting probe. *J Biomed Opt 10,* 054010.
94. Dafni, H., Cohen, B., Ziv, K., Israely, T., Goldshmidt, O., Nevo, N., Harmelin, A., Vlodavsky, I., & Neeman, M. (2005). The role of heparanase in lymph node metastatic dissemination: dynamic contrast-enhanced MRI of Eb lymphoma in mice. *Neoplasia 7,* 224–33.
95. Talanov, V. S., Regino, C. A., Kobayashi, H., Bernardo, M., Choyke, P. L., & Brechbiel, M. W. (2006). Dendrimer based nanoprobe for dual modality magnetic resonance and fluorescence imaging. *Nano Lett 6,* 1459–63.
96. Koyama, Y., Talanov, V. S., Bernardo, M., Hama, Y., Regino, C. A., Brechbiel, M. W., Choyke, P. L., & Kobayashi, H. (2007). A dendrimer-based nanosized contrast agent dual-labeled for magnetic resonance and optical fluorescence imaging to localize the sentinel lymph node in mice. *J Magn Reson Imaging 25,* 866–71.
97. Kobayashi, H., Koyama, Y., Barrett, T., Hama, Y., Regino, C. A. S., Shin, I. S., Jang, B. S., Le, N., Paik, C. H., Choyke, P. L., & Urano, Y. (2007). Multimodal nanoprobes for radionuclide and five-color near-infrared optical lymphatic imaging. *Acs Nano 1,* 258–264.
98. Bumb, A., Brechbiel, M. W., & Choyke, P. (2010). Macromolecular and dendrimer based magnetic resonance contrast agents. *Acta Radiol. 51*(7), 751–767.

CHAPTER 9

MISCELLANEOUS BIOMEDICAL APPLICATIONS OF DENDRIMERS

ZAHOOR AHMAD PARRY, PhD, and RAJESH PANDEY, MD

CONTENTS

9.1 DENDRITIC SENSORS

Although dendrimers are single molecules, they can contain large numbers of important functional groups on their surface. This makes them highly suitable for applications where the covalent interaction or close proximity of a large number of chemical species is important. Researchers have investigated the fluorescence of a G4 poly-propylene amine dendrimer bearing 32 dansyl units in the periphery (Figure 9.1) [1]. Because the dendrimer contains thirty aliphatic amine moieties in the interior, suitably selected metal ions would be able to coordinate. It was observed that when CO_2^+ ion was incorporated into the structure of the dendrimer, the powerful fluorescence of every dansyl unit was quenched. Very low concentrations of CO_2^+, could be detected by using a dendrimer concentration

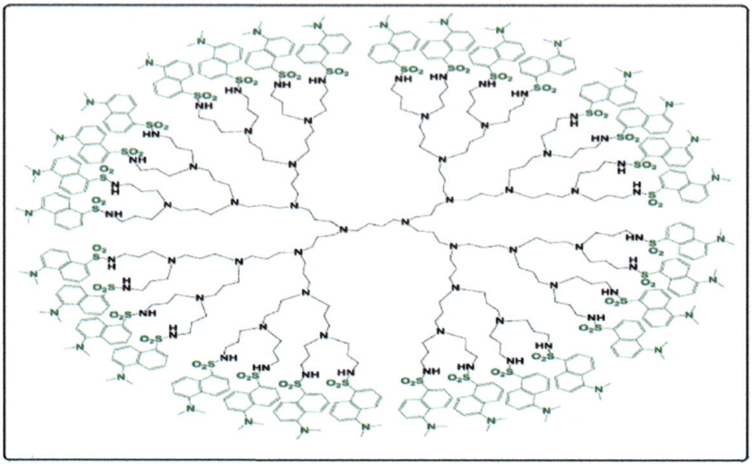

FIGURE 9.1 32 peripheral dansyl units in a poly (propylene amine) dendrimer. (© Abbasi et al.; licensee Springer. 2014. Reprinted with permission via the Creative Commons Attribution License. From Elham Abbasi, Sedigheh Fekri Aval, Abolfazl Akbarzadeh, Morteza Milani, Hamid Tayefi Nasrabadi, Sang Woo Joo, Younes Hanifehpour, Kazem Nejati-Koshki and Roghiyeh Pashaei-Asl. Dendrimers: synthesis, applications, and properties, Nanoscale Research Letters 2014, 9:247.)

of 4.6×10^{-6} M. Conceptually, the multiple fluorescent moieties on the surface helped to amplify the sensitivity of the polymer as a biosensor [2].

9.2 BIOSENSORS

Biosensors are specific analytical tools which can be used for determining the concentrations of various biomolecules, including ions, in biological fluids. They are often designed using electrodes/optical transducers that are coupled with catalytic and/or molecular recognition elements which may be enzymes and/or antibodies. Consequently, immobilization of proteins on the surface of transducers but without any loss of the biological activity is a crucial issue in the development of biosensors. For this purpose, layer-by-layer (LbL) dendrimer films containing proteins have been extensively studied in order to improve the performance of biosensors [3–7]. In this context, the important role of dendrimers in biosensor designing can be envisaged, and includes:

1. Formation of surface monolayers acting as scaffolds for immobilizing proteins.

2. Depositing the multilayer LbL films constituted of dendrimers and proteins.
3. Encapsulation or covalent attachment of metal species and mediators of electron transfer.

Several research groups have used dendrimers so as to modify the surface of electrodes and/or other devices with monolayer films. On these films, proteins may be immobilized either covalently or non-covalently. Mono-molecular layers of PAMAM dendrimers have been prepared on the surface of metal substrates such as Au or Ag for use as surface plasmon sensors [8–10]. The PAMAM monolayers were modified further using proteins or single–stranded DNA (ssDNA) in order to fabricate the biosensors. In these examples, the primary amines in PAMAM dendrimer had been covalently linked with the surface of the substrate and the biomolecules. Other research groups have reported the development of electrochemical biosensors that were prepared using electrodes modified with a monolayer of dendrimer coupled with an aptamer as well as nanoclusters of Pt [11, 12]. Obviously, the clearly-defined and compact conformation of dendrimers is advantageous in the designing of biosensor interfaces.

Many research groups have studied the formulation of dendrimer/enzyme multilayer films and their implications for developing biosensors. LbL films made up of PAMAM dendrimers and glucose oxidase were coated on the surface of Au electrodes to develop glucose biosensors. The PAMAM layers were covalently linked with glucose oxidase via Schiff's base linkages [13]. The amperage (output current) of the glucose sensors increased linearly with the increasing number of PAMAM/glucose oxidase bilayers, thereby confirming that the catalytic activity of glucose oxidase could be preserved in the LbL film. These results indicate that glucose could smoothly permeate the porous PAMAM/glucose oxidase multilayer films. In a somewhat similar protocol, ferrocene–tethered PAMAM was used to construct reagent-free glucose sensors that may be used without the need to add any electron transfer agent to the sample solution [14]. Another research group demonstrated the potential use of LbL films comprised of Au nanoparticles and ferrocene-tethered PAMAM for developing amperometric biosensors [15]. PAMAM/glucose oxidase LbL films had been prepared through electrostatic linkage between an amine–terminated PAMAM dendrimer and glucose oxidase (negatively charged) at neutral pH [16]. Just two bilayer films of PAMAM/glucose oxidase could be prepared. However any further deposition of glucose oxidase layers failed to enhance the output current of the glucose sensors. Thus, the covalently bonded PAMAM/glucose oxidase films were certainly superior to the electrostatic bonding films as far as constructing glucose biosensors was concerned. The compact, globular conformation of PAMAM might be somewhat less effective in forming the complementary electrostatic bonds with glucose oxidase and this is in clear contrast to the successful bonding of linear polyamines as well as lectin to glucose oxidase [17, 18]. On the other hand, the LbL films comprised of Cl-catechol-1,2-dioxigenase and PAMAM dendrimer were stable although they were prepared through electrostatic bonding [19]. Other researchers reported the construction of electrostatic bonded LbL films of hemoproteins and PPI dendrimer under various pH conditions [20]. Interestingly, it was noted that hemoglobin (Hb)/PPI multilayer films can be successfully assembled at pH 9.0 (at which Hb is negatively charged) besides at pH 5.0 (at which Hb as well as PPI are positively charged). At pH 5.0, the LbL film formation was attributed to the localized electrostatic attractions or the charge reversal of Hb that was induced on the PPI surface. The results indicate that the stability of electrostatic binding in the dendrimer/protein films depends to a large extent on the type of protein. This could be possibly due to the different charge distributions on the protein surface.

Metal nanoparticles have been successfully encapsulated into dendrimers [21]. For example, Pt and Au nanoparticle-encapsulating dendrimers were widely used to construct the enzyme LbL films for applications in biosensors. Researchers have prepared LbL films comprising Au nanoparticle-encapsulating PPI dendrimer/myoglobin. The films were applied on the surface of graphite electrodes and the catalytic response of these electrodes was higher when compared with the electrodes modified with LbL films but lacking Au nanoparticles [22]. Likewise, PAMAM dendrimers encapsulating Pt or CdS semiconductor have been also used for preparing glucose biosensors [23, 24]. In another work, Pt or Au-encapsulating dendrimers had been coupled with enzymes and carbon nanotubes in LbL films in order to construct biosensors that would be sensitive to pesticides and glutamate [25–27]. The modifications of dendrimer LbL films using metal hexa-cyanoferrate nanoparticles was shown to be effective for favorably enhancing the electrochemical response of biosensors to compounds like hydrogen peroxide and glucose [28–30].

Of late, dendrimer LbL films have also been employed as gating materials for preparing FET (field effect transistor) biosensors. Extensive work showed the preparation of LbL films with alternating deposition of special carboxylated single-walled carbon nanotubes and PAMAM dendrimers, on the surface of an FET gate to develop penicillin biosensors [31, 32]. The gate potential of FET biosensor was found to be sensitive to penicillin in the concentration ranging from 5.0×10^{-6} M to 2.5×10^{-2} M. Other workers have used LbL films comprised of TsPc (tetra-sulfonated phthalocyanine) and dendrimers for constructing FET sensors that would be sensitive to various stimuli such as pH, humidity, and glucose [33–35]. The results documenting the high performance of FET sensors were attributed to the porous structure of the LbLTsPc/dendrimer films that were permeable to H^+ and glucose.

9.3 DENDRIMERS USED FOR ENHANCING SOLUBILITY

PAMAM dendrimers are expected to possess potential applications to enhance the solubility for various drug delivery systems. Dendrimers are known to have hydrophilic exteriors as well as interiors, and this is responsible for its unimolecular, micelle-like nature. Dendrimer-based carriers provide an opportunity to enhance the oral bioavailability of many problematic drugs. Thus, dendrimer-based nano-formulations offer the potential for enhancing the bioavailability of those drugs that are otherwise poorly soluble, besides those that are substrates for efflux transporters [36, 37].

9.4 PHOTODYNAMIC THERAPY

PDT (photodynamic therapy) depends on the activation of a particular photosensitizing agent with either visible or near-infrared (NIR) radiation. Following the excitation process, a highly energetic state is formed which, subsequent to its reaction with oxygen, generates a highly reactive singlet oxygen which is capable of inducing apoptosis or necrosis in neoplastic cells. The dendrimer-based delivery of PDT agents has been widely investigated in the past few years so as to improve their tumor selectivity, tissue retention, pharmacokinetics and pharmacodynamics [38–41].

LbL film-coated iron oxide (Fe_3O_4) nanoparticles have been prepared with the intention of targeting and imaging neoplastic cells. The surface of Fe_3O_4 nanoparticles was coated with LbL layers of polylysine (PL)/polyglutamic acid (PGA) and followed by cross-linking with carbodiimide reagent. Further, the outermost surface of LbL layer was modified with folic acid (FA)-linked PAMAM [42]. The crosslinking of the PL/PGA layers was essential to stabilize the modified Fe_3O_4 nanoparticles, which could successfully target the FA receptors that are known to be overexpressed on the surface of tumor cells [43]. It was observed that the FA-modified Fe_3O_4 nanoparticles targeted the neoplastic cells at a very small volume (0.60 ± 0.15 cm^3). In addition, they could be successfully employed for magnetic resonance imaging of the cells.

The adsorption of cells/proteins on the surface of LbL film-coated PAMAM dendrimers has been studied for evaluating the biocompatibility of these films. PAMAM/PSS LbL films layered on the surface of flat substrates/microparticles were had been further modified using polyethylene glycol (PEG)-anchored lipids in order to prevent the adsorption of human serum albumin (HSA) as well as the macrophage cell line [44]. The PEG-modified surface of LbL films was observed to be resistant to cell adhesion/HSA adsorption when compared to the surface of unmodified cationic PAMAM/PSS films. In a different study on hydrazine phosphorus dendrimer LbL films, the influence of surface charges of on the adhesion/maturation of fetal cortical rat neurons was assessed. The neurons were found to adhere firmly and mature faster on the LbL film surfaces bearing positive charges compared to those with negative charges [45].

9.5 NOW AND NEXT?

There are numerous other areas of biological chemistry in which the applications of dendrimer-based systems may be of help although the evidence, for some of them, is either scanty at the moment or a wide bridge still exists between basic research and its clinical translation (Table 9.1).

TABLE 9.1 Emerging and Upcoming Biomedical Applications of Dendrimers

- Purification of dendrimer-based commercial products (e.g., cosmetics) contaminated by toxic metal ions and organic/inorganic solutes [46].
- Developing highly sensitive analytical devices [47, 48].
- Prion research [49].
- Burn treatment [50].
- EPR imaging with spin-labeled dendrimers [51, 52].

KEYWORDS

- **biosensors**
- **carbon nanotubes**
- **enhancing solubility**
- **nano-formulations**
- **nanoclusters**
- **photodynamic therapy**

REFERENCES

1. Balzani, V., Ceroni, P., Gestermann, S., Kauffmann, C., Gorka, M., & Vögtle, F. (2000). Dendrimers as fluorescent sensors with signal amplification. *Chem Commun 2000*, 853–854.
2. Beer, P. D., Gale, P. A., & Smith, D. K. (1999). Supramolecular Chemistry. Oxford: Oxford University Press.
3. Iost, R. M., & Crespilho, F. N. (2012). Layer-by-layer self-assembly and electrochemistry: Applications in biosensing and bioelectronics. *Biosens. Bioelectron. 31*, 1–10.
4. Shi, H., Yang, Y., Huang, J., Zhao, Z., Xu, X., Anzai, J., Osa, T., & Chen, Q. (2006). Amperometric choline biosensors prepared by layer-by-layer deposition of choline oxidase on the Prussian blue-modified platinum electrode. *Talanta 70*, 852–858.
5. Zhao, W., Xu, J.-J., & Chen, H.-Y. (2006). Electrochemical biosensors based on layer-by-layer assemblies. *Electroanalysis 18*, 1737–1748.
6. Siqueira, J. R., Jr., Caseli, L., Crespilho, F. N., Zucolotto, V., & Oliveira, O. N., Jr. (2010). Immobilization of biomolecules on nanostructured films for biosensing. *Biosens. Bioelectron. 23*, 1254–1263.

7. Huang, J., Yang, Y., Shi, H., Song, Z., Zhao, Z., Anzai, J., Osa, T., & Chen, Q. (2006). Multi-walled carbon nanotubes-based glucose biosensor prepared by a layer-by-layer technique. *Mater. Sci. Eng. C 26,* 113–117.
8. Mark, S. S., Sandhyarani, N., Zhu, C., Campagnolo, C., & Batt, C. A. (2004). Dendrimer-functionalized self-assembled monolayers as a surface plasmon resonance sensor surface. *Langmuir 20*, 6808–6817.
9. Sihgh, P., Onodera, T., Mizuta, Y., Matsumono, K., Miura, N., & Toko, K. (2009). Dendrimer modified biochip for detection of 2,3,6-trinitrotoluene on SPR immunosensor: Fabrication and advantages. *Sens. Actuators B 137*, 403–409.
10. Feng, C. L., Yin, M., Zhang, D., Zhu, S., Caminade, A. M., Majoral, J. P., & Müllen, K. (2011). Fluorescent core–shell star polymers based biosensors for ultrasensitive DNA detection by surface plasmon fluorescence spectroscopy. *Macromol. Rapid Commun. 32*, 679–683.
11. Qian, L., Liu, Y., Song, Y., Li, Z., & Yang, X. (2005). Electrodeposition of Pt nanoclusters on the surface modified by monolayer poly(amidoamine) dendrimer film. *Electrochem. Commun. 7,* 1209–1212.
12. Zhang, Z., Yang, W., Wang, J., Yang, C., Yang, F., & Yang, X. (2009). A sensitive impedimetric thrombin aptasensor based on polyamidoamine dendrimer. *Talanta 78*, 1240–1245.
13. Yoon, H. C., & Kim, H.-S. (2000). Multilayered assembly of dendrimers with enzymes on gold: Thickness-controlled biosensing interface. *Anal. Chem. 72*, 922–926.
14. Yoon, H. C., Hong, M.-Y., & Kim, H.-S. (2000). Functionalization of a poly(amidoamine) dendrimer with ferrocenyls and its application to the construction of a reagentless enzyme electrode. *Anal. Chem. 72,* 4420–4427.
15. Suk, J., Lee, J., & Kwak, J. Electrochemistry on alternate structures of gold nanoparticles and ferrocene-tethered polyamidoamine dendrimers. Bull. Korean Chem. Soc. (2004). 25, 1681–1686.
16. Lojou, E., & Bianco, P. (2006). Assemblies of dendrimers and proteins on carbon and gold electrodes. *Bioelectrochem. 69*, 237–247.
17. Hodak, J., Etchenique, R., & Calvo, E. J. (1997). Layer-by-layer self-assembly of glucose oxidase with a poly (allylamine) ferrocene redox mediator. *Langmuir 13*, 2708–2716.
18. Anzai, J., Kobayashi, Y. (2000). Construction of multilayer thin films of enzymes by means of sugar-lectin interactions. *Langmuir 16,* 2851–2856.
19. Zucolotto, V., Pinto, A. P. A., Tumolo, T., Moreas, M. L., Baptista, M. S., Riul, A., Jr., Araujo, A. P. U., & Oliveira, O. N., Jr. (2006). Catechol biosensing using a nanostructured layer-by-layer film containing Cl-catechol 1,2-dioxygenase. *Biosens. Bioelectron. 21*, 1320–1326.
20. He, P., Li, M., & Hu, N. (2005). Interaction of heme proteins with poly (propyleneimine) dendrimers in layer-by-layer assembly films under different pH conditions. *Biopolymers 79*, 310–323.
21. Scott, R. W. J., Wilson, O. M., & Crooks, R. M. (2005). Synthesis, characterization, and applications of dendrimer-encapsulated nanoparticles. *J. Phys. Chem. B 109,* 692–704.
22. Zhang, H., & Hu, N. (2007). Assembly of myoglobin layer-by-layer films with poly(propyleneimine) dendrimer-stabilized gold nanoparticles and its application in electrochemical biosensing. *Biosens. Bioelectron. 23*, 393–399.

23. Sun, J., Zhu, Y., Yang, X., & Li, C. (2009). Photoelectrochemical glucose biosensor incorporating CdS nanoparticles. *Particulology 7,* 347–352.
24. Zhu, Y., Zhu, H., Yang, X., Xu, L., & Li, C. (2007). Sensitive biosensors based on (dendrimer encapsulated Pt nanoparticles)/enzyme multilayers. *Electroanalysis 19,* 698–703.
25. Qu, Y., Sun, Q., Xiao, F., Shi, G., & Jin, L. (2010). Layer-by-layer self-assembled acetylcholineesterase/PAMAM-Au on CNTs modified electrode for sensing pesticides. *Bioelectrochemistry 77,* 139–144.
26. Tang, L., Zhu, Y., Yang, X., & Li, C. (2007). An enhanced biosensor for glutamate based on self-assembled carbon nanotubes and dendrimer-encapsulated platinum nanobiocomposites–doped polypyrrole film. *Anal. Chim. Acta 597,* 145–150.
27. Tang, L., Zhu, Y., Xu, L., Yang, X., & Li, C. (2007). Amperometric glutamate biosensor based on self-assembling glutamate dehydrogenase and dendrimer–encapsulated platinum nanoparticles onto carbon nanotubes. *Talanta 73,* 438–443.
28. Crespilho, F. N., Ghica, M. E., Zucolotto, V., Nart, F. C., Oliveira, O. N., Jr., & Brett, C. M. A. (2007). Electroactive nanostructured membranes (ENM): Synthesis and electrochemical properties of redox mediator-modified gold nanoparticles using a dendrimer layer-by-layer approach. *Electroanalysis 19,* 805–812.
29. Crespilho, F. N., Ghica, M. E., Gouveia-Caridale, C., Oliveira, O. N., Jr., & Brett, C. M. A. (2008). Enzyme immobilization on electroactive nanostructured membranes (ENM): Optimised architectures for biosensing. *Talanta 76,* 922–928.
30. Crespilho, F. N., Ghica, M. E., Florescu, M., Nart, F. C., Oliveira, O. N., Jr., & Brett, C. M. A. (2006). A strategy for enzyme immobilization on layer-by-layer dendrimer-gold nanoparticle electrocatalytic membrane incorporating redox mediator. *Electrochem. Commun. 8,* 1665–1670.
31. Siqueira, J. R., Jr., Abouzar, M. H., Poghossian, A., Zucolotto, V., Oliveira, O. N., Jr., & Schöning, M. J. (2009). Penicillin biosensor based on a capacitive field-effect structure functionalized with a dendrimer/carbon nanotube multilayer. *Biosens. Bioelectron. 25,* 497–501.
32. Siqueira, J. R. Jr., Werner, C. F., Bäcker, M., Poghossian, A., Zucolotto, V., Oliveira, O. N., Jr., Schöning, M. J. (2009). Layer-by-layer assembly of carbon nanotubes incorporated in light-addressable potentiometric sensors. *J. Phys. Chem. C 113,* 14765–14770.
33. Fernandes, E. G. R., Vieira, N. C. S., de Queiroz, A. A. A., Guimarães, F. E. G., & Zucolotto, V. (2010). Immobilization of poly(propylene imine) dendrimer/nickel phthalocyanine as nanostructured multilayer films to be used as gate membranes for SEGFET pH sensors. *J. Phys. Chem. C 114,* 6478–6483.
34. Centurion, L. M. P. C., Moreira, W. C., & Zucolotto, V. (2012). Tailoring molecular architectures with cobalt tetrasulfonated phthalocyanine: Immobilization in layer-by-layer films and sensing applications. *J. Nanosci. Nanotechnol. 12,* 2399–2405.
35. Vieira, N. C. S., Figueiredo, A., Queiroz, A. A. A., Zucolotto, V., & Guimarãez, F. E. G. (2011). Self-assembled films of dendrimers and metallophthalocyanines as FET–based glucose biosensors. *Sensors 11,* 9442–9449.
36. Tomalia, D. A., Baker, H., Dewald, J. R., Hall, M., Kallos, G., Martin, S., Roeck, J., Ryder, J., Smith, P., & Dendrimers, I. I. (1986). Architecture, nanostructure and supramolecular chemistry. *Macromolecules 19,* 2466.

37. Froehling, P. E. (2001). Dendrimers and dyes – a review. *Dyes Pigments 48*, 187–195.
38. Triesscheijn, M., Baas, P., Schellens, J. H., & Stewart, F. A. (2006). Photodynamic therapy in oncology. *Oncologist 11*, 1034–1044.
39. Nishiyama, N., Stapert, H. R., Zhang, G. D., Takasu, D., Jiang, D. L., Nagano, T., Aida, T., & Kataoka, K. (2003). Light-harvesting ionic dendrimer porphyrins as new photosensitizers for photodynamic therapy. *Bioconjug Chem 14*, 58–66.
40. Zhang, G. D., Harada, A., Nishiyama, N., Jiang, D. L., Koyama, H., Aida, T., & Kataoka, K. (2003). Polyion complex micelles entrapping cationic dendrimer porphyrin: effective photosensitizer for photodynamic therapy of cancer. *J Control Release 93,* 141–150.
41. Battah, S. H., Chee, C. E., Nakanishi, H., Gerscher, S., MacRobert, A. J., & Edwards, C. (2001). Synthesis and biological studies of 5-aminolevulinic acid containing dendrimers for photodynamic therapy. *Bioconjug Chem 12,* 980–988.
42. Shi, X., Wang, S. H., Swanson, S. D., Ge, S., Cao, Z., Van Antwerp, M. E., Landmark, K. L., & Baker, Jr., J. R. (2008). Dendrimer-functionalized shell-crosslinked iron oxide nanoparticles for in-vivo magnetic resonance imaging of tumors. *Adv. Mater. 20,* 1671–1678.
43. Wang, S. H., Shi, X., Van Antwerp, M., Cao, Z., Swanson, S. D., Bi, X., & Baker, J. R., Jr. (2007). Dendrimer-functionalized iron oxide nanoparticles for specific targeting and imaging of cancer cells. *Adv. Funct. Mater. 17,* 3043–3050.
44. Khopade, A. J., & Caruso, F. (2003). Surface-modification of polyelectrolyte multilayer-coated particles for biological applications. *Langmuir 19*, 6219–6225.
45. Hernandez-Lopez, J.-L., Khor, H. L., Caminade, A.-M., Majoral, J.-P., Mittler, S., Knoll, W., & Kim, D. H. (2008). Bioactive multilayer thin films of charged, N., N-di-substituted hydrazine phosphorus dendrimers fabricated by layer-by-layer self-assembly. *Thin Solid Films 516*, 1256–1264.
46. Tiwari, D. K., Behari, J., & Sen, P. (2008). Application of nanoparticles in waste water treatment. *World Appl Sci J 3*, 417–433.
47. Yoon, H. C., Lee, D., & Kim, H-S. (2002). Reversible affinity interactions of antibody molecules at functionalized dendrimer monolayer: affinity-sensing surface with reusability. *Anal Chim Acta 456,* 209–218.
48. Benters, R., Niemeyer, C. M., Drutschmann, D., Blohm, D., & Wohrle, D. (2002). DNA microarrays with PAMAM dendritic linker systems. *Nucleic Acid Res 30*, 1–11.
49. McCarthy, J. M., Appelhans, D., Tatzelt, J., & Rogers, M. S. (2013). Nanomedicine for prion disease treatment. New insights into the role of dendrimers. *Prion 7*(3), 198–202.
50. Halkes, S. B. A., Vrasidas, I., Rooijer, G. R., van den Berg, A. J. J., Liskamp, R. M. J., & Pieters, R. J. (2002). Synthesis and biological activity of polygalloyl-dendrimers as stable tannic acid mimics. *Bioorg Med Chem Lett 12*, 1567–1570.
51. Yordanov, A. T., Yamada, K.-I., Krishna, M. C., Mitchell, J. B., Woller, E., Cloninger, M., & Brechbiel, M. W. (2001). Spin-labeled dendrimers in EPR imaging with low molecular weight nitroxides. *Angew Chem Int Ed Engl 40*, 2690–2692.
52. Abolfazl, A., Samiei, M., & Soodabeh, D. (2012). Magnetic nanoparticles: preparation, physical properties and applications in biomedicine. *Nanoscale Res Lett 7*, 144–157.

CHAPTER 10

TOXICITY OF DENDRIMERS

ZAHOOR AHMAD PARRY, PhD, and RAJESH PANDEY, MD

CONTENTS

10.1 INTRODUCTION

Every year thousands of papers describing the merits of using PAMAM dendrimers in various areas of science/technology get published. Nevertheless, one should bear in mind that there are some demerits limiting the use of these dendrimers. Interestingly, the potential toxicity of dendrimers has surfaced after nearly 3 decades of research on dendrimers. The key aspects that have emerged as of top priority to be resolved so as to successfully employ dendrimers in nanotechnology are [1–3]:

1. Significant reduction of toxicity through appropriate terminal groups modifications.
2. The relevance of the outcomes from the in vitro and in vivo studies (regarding dendrimers' toxicity).
3. A clear-cut understanding of the interactions between dendrimers and blood components including the coagulation pathway and the fibrinolytic system.

4. Understanding immune responses to dendrimers' as well as the consequences of such interactions.
5. Understanding the sequence of alterations in endothelial cell functioning to assess the effect of dendrimers on blood vessels.
6. Development of dendrimer-based drug nano-formulations with low clearances, short plasma half-lives, and well-understood intravascular degradation profiles.
7. Knowledge of the most efficient and safest routes of administration.

Well-documented scientific knowledge of PAMAMs' toxicity in an organism is available concerning either long- or short-term applications of PAMAM. An oral acute toxicity study lasted nearly ten days, during which CD-1 mice were given a cationic dendrimer (PAMAM G4) at a dose of 50 mg/kg body weight. The maximum tolerated dose of was established as a function of size as well as surface functionality, and clinical signs of toxicity were monitored. Observations with longer periods of dendrimer administration were conducted by other authors [1] who studied acute, subacute, and chronic toxicity in male Swiss–Webster mice for 7 days, 30 days, and 6 months. Animals were monitored for day-to-day behavioral abnormalities and changes in body weight. Many reports concerning the *in vivo* use of PAMAMs in animal models (mice/rats) tend to use higher doses for shorter periods of time, from 2 hours to a few weeks [5–9]. It is likely that the high cost of dendrimers (which strongly influences the overall cost of a study) might have significantly impacted the experimental designs of the studies and supported the rationales of short-term protocols. More importantly though at least considering the economical aspect, it needs to be stressed that dendrimers are currently expensive to an extent that they discourage long-term experimental designs, so much so that their potential for widespread use in medical practice seems more an illusion than reality. However, at least with regard to using PAMAM dendrimers for chronic diseases (e.g., as hypoglycemic agents), one might rather reasonably consider using expensive compounds for a long-term treatment, keeping in mind that the therapy of people suffering from chronic disease (e.g., diabetes mellitus) is life-long for all practical purposes [10].

In July 2003, the FDA permitted the first clinical trials of a dendrimer-based pharmaceutical (Vivagel™, a topical microbicide for the prevention of HIV infection in women) developed by an Australian company,

Starpharma. Another dendrimer undergoing preclinical evaluation is the multi-antigenic peptide PHSCN-lysine dendrimer, employed in a metastatic murine cancer model in an attempt to inhibit the invasion/growth of breast cancer cells via integrin (an adhesion molecule) targeting [11]. In 2006, the FDA launched a Nanotechnology Task Force for critical regulatory matters concerning nanomaterials containing drug formulations. Accordingly, guidelines for IND (*Investigational New Drug Application*) are being developed to address specific issues of nanomaterials containing drug formulations, which have physicochemical properties distinct from drug molecules. This highlights the importance of biodistribution and pharmacokinetics as a topic of the assessment of absorption, distribution, metabolism, and excretion (ADME). Moreover, the FDA recommends a long-term planned study as a standard guideline to assess their safety as well as efficacy. As reported, dendrimers such as PAMAM are well tolerated in mice without any serious toxicity [12, 13]. However, some vacuolization of the cytoplasm in the liver has been observed after long-term administration of cationic PAMAM dendrimers (G3–7) at 2.5–10 mg/kg body weight [9]. It is important to consider not only the properties of the dendrimer or dendrimer-based drug formulation, but also the experimental design itself, for careful evaluation and interpretation of ADME with regard to safety and efficacy during a possible clinical use.

Another issue which needs to be underscored is that a considerable inconsistency and variability of outcomes originating from animal studies may be simply because of the diversity of the used animal models and strains, differentiated treatment durations, and seasonal fluctuations. It has been noticed and emphasized that while working with animals, the phenomenon of seasonal fluctuations should be remembered because such fluctuations can potentially affect animal metabolism and bioenergetics, despite maintaining the animals under a standard housing regime [14–19]. Another important issue concerns the use of different laboratory rodent strains to investigate similar models of pathology. Some strains may demonstrate altered sensitivity to the tested drugs or may undergo better adaptation to their possibly harmful effects [14–17, 20]. Consequently, the outcomes gathered from studies conducted on different strains of animals, or even on animals provided by multiple distributors, may differ considerably, which in turn, may influence data interpretation and laying of conclusions and guidelines [21]. Overall, testing of the pharmacological or toxicity profiles in laboratory animals are associated with some confounders, such as seasonality or animal strain and origin, that are more than likely to significantly alter the perception of true facts.

As the biological safety of unmodified PAMAM dendrimers has been questioned, efforts are on the way to modify dendrimer surface groups by conjugating them with specific molecules such as fatty acids, carbohydrates, or polyethylene glycol (PEG) [22]. Out of these, conjugation with PEG (PEGylation) has been considered as the method of choice for reducing PAMAM toxicity besides enhancing the biocompatibility of the dendrimers. This is mainly because PEG is nontoxic, non-immunogenic, and has favorable pharmacokinetics, half-life as well as biodistribution [23]. In particular, the *in vitro* experiments performed with using various human/animal cell lines have shown that increasing attachment with PEG of amino-terminated PAMAM results in markedly decreased cytotoxicity of these dendrimers [24, 25]. Moreover, it has been observed that whereas plain PAMAMs are able to stimulate the overproduction of reactive oxygen species (ROS) which altered the functioning of mitochondria, even leading to cell death, their PEGylated counterparts inhibited these phenomena [26, 27].

When the strategy of dendrimer surface modifications (e.g., PEGylation) became more familiar and began to gain foothold, other modifying chemicals began to be used. Thus, it was noted that covalent attachment of acetyl [28] and lauroyl [22] groups to the peripheral amino groups of PAMAM dendrimers decreased the cytotoxicity. This is probably because of the reduction of the number of available protonated amino groups as well as shielding of the positive charges on the dendrimers' surface. It is worth mentioning that PAMAM acetylation has been found to increase the ability of delivering siRNA to tumor cells [29]. Other efforts to modify PAMAM dendrimers for gene/siRNA delivery have also met with success as far as decreasing dendrimer cytotoxicity and increasing the efficiency of DNA delivery are concerned [30]. Theses modified dendrimers have been effectively used in several nanomedical applications as discussed in the previous chapters of this book.

Thus, it is clear from the above discussion that surface modification of dendrimers using the correct chemical group and to the correct extent (i.e., degree of modification) not only bears the advantage of improved stereo-geometry and spatial structure, but also achieves considerably reduced toxicity. In other words, the biological profile of a modified dendrimer is expected to be quite different from that of an unmodified dendrimer. Reports suggest that a fourth generation PAMAM dendrimer modified by pyrrolidone residues demonstrates no major hemolytic activity, showing minor cytotoxic effects, as compared with its parent counterpart [31]. In addition, its reactivity as well as ability to bind to human serum albumin also became reduced, compared to the unmodified PAMAM G4.

Thus, considering the title of this chapter, we are faced with a choice of modified versus unmodified dendrimers. On one hand, based on available evidence accumulated over the past few years, one might anticipate that unmodified PAMAM dendrimers should probably not be unrestrainedly considered as safe drug formulations to be directly used in the treatment of chronic disorders. Their unpredictable biological activity, as well as their high potential to trigger nonspecific, often difficult to control, and toxic side effects should be a boon to researchers to clearly verify/validate the various possible mechanisms of their toxicity.

10.2 GLYCO-PAMAM DENDRIMERS AND GLYCOSYLATED PAMAM DENDRIMERS

In recent years, it has become clear that carbohydrates play an important role in various biological processes [32]. Because of the high biocompatibility and biodegradability, carbohydrate-based formulations have found wide pharmaceutical and other medical applications. It is further apparent that synthetic carbohydrate polymers (the so-called "glyco-polymers") can exhibit specific interactions with proteins. Thus, one can assume that a multitude of biological interactions of natural carbohydrates can be mimicked by synthetic glyco-polymers containing specific sugar moieties [33] (Figure 10.1). The first glyco-dendrimers [34] were reported in 1993, and since then this field of glyco-dendrimers has evolved, matured and proliferated to an unprecedented level. It is worth mentioning that glyco-dendrimers were initially designed as bioisosteres of cell surface multi-antennary molecules, but very soon polymer scientists recognized

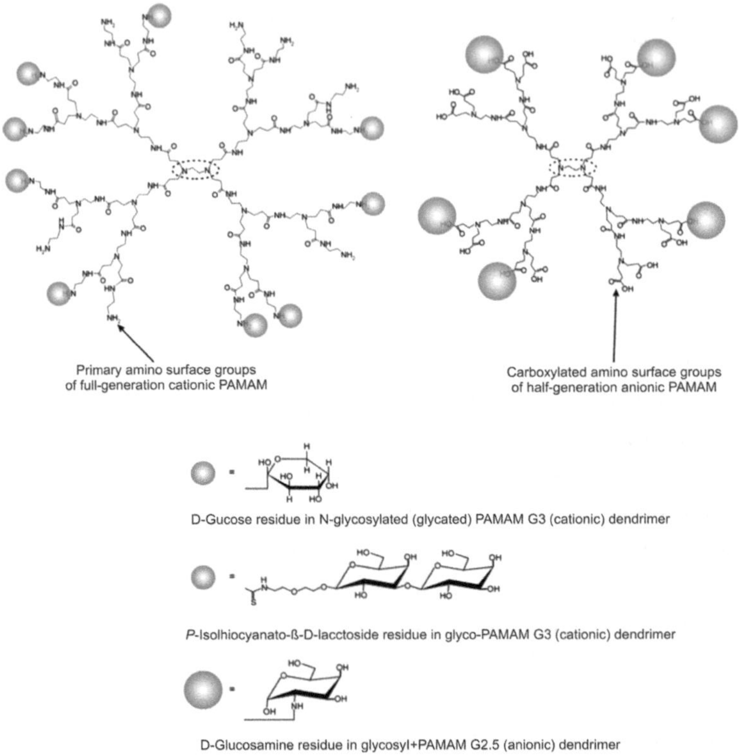

FIGURE 10.1 Glyco-dendrimers, glycosylated dendrimers, and N-glycosylated (glycated) PAMAM dendrimers. Surface functional groups can be modified with various residues (solid balls), e.g., mono- or oligosaccharides. (From: Magdalena Labieniec-Watala, Cezary Watala. PAMAM Dendrimers: Destined for Success or Doomed to Fail? Plain and Modified PAMAM Dendrimers in the Context of Biomedical Applications. Journal of Pharmaceutical Sciences 104:2–14, 2015; DOI10.1002/jps.24222. http://onlinelibrary.wiley.com/doi/10.1002/jps.24222/full)

and became inspired by the potential biomedical applications of this class of dendrimers. This, in turn, triggered further experiments, investigations and the slow but steady development of new synthetic strategic finally resulting in these nanostructures [35]. Since then, one has witnessed considerable progress especially in fields of highly advanced chemistry, e.g., organo-metallic chemistry, chemo-enzymatic catalysis, silicon chemistry, and dendrimer self-assembly to say the least [36].

Given their commercial potential, PAMAMs have been the most widely used scaffolds in current investigations involving dendrimers. PAMAM-based dendrimers, having in-built amine functionalities on their surfaces, are the first and most commonly used scaffolds for carbohydrate attachment. They have been modified with a wide variety of sugars or sugar derivatives and with variable sugar densities. Scaffolded glyco-dendrimers are non-immunogenic, a very important property if they are to be used as bacterial or viral anti-adhesins [37]. The first example [38] of saccharide-substituted PAMAM dendrimer synthesis was that of the so-called "sugar balls" (via an amide-bond formation beginning with sugar lactones). This process suffered from the disadvantage of relinquising the reducing sugars that also served as extended linkers [39]. Subsequently, several other strategies have been described to functionalize PAMAM dendrimers with carbohydrates [22, 35, 40–44]:

1. Introduction of thiourea linkages, formed by reacting amino-dendrimers with isothiocyanated carbohydrate derivatives.
2. Direct amide linkages with carbohydrates-bearing carboxylated/activated ester derivatives.
3. Reductive amination.
4. Incorporation of chloro-/bromo-acetamide groups on PAMAM dendrimers or saccharides to provide highly reactive electrophiles.
5. Saccharide units can be localized in the center of the dendrimer (part of its core) building the dendrimer side branches or covering the outer surface after attachment to terminal groups.

A diverse array of carbohydrates (Table 10.1) have been employed to transform the structure of a dendrimer in order to suit the needs of a researcher, and to offer the potential for its use in equally diverse applications [45–49].

TABLE 10.1 Carbohydrates Used to Alter Dendrimer Structure and Functionality

• Chitosan.
• Carboxymethyl chitosan.
• Heparin.
• Galactose.
• Glucose.
• Mannose.
• Lactose.

Besides the glyco-dendrimers, another class of dendrimers, namely "glycosylated dendrimers" has been delineated. In contrast to glyco-dendrimers, glycosylated dendrimers are formed from the anionic PAMAM dendrimers (half-generation PAMAMs). Their surfaces are modified with glucosamine units [50] (Figure 10.1). Nevertheless, both the terms involve similar groups of compounds, are characterized by similar properties, and are synthesized to obtain dendrimers which are better suited for various biomedical applications compared with their unmodified counterparts. Noteworthy, both glyco-dendrimers and glycosylated dendrimers (irrespective of whether they are generated during designed chemical synthesis or spontaneously, in the presence of surplus glucose), have one common characteristic. They are shielded by sugar units, making them more hydrophilic and less cytotoxic. Hence, they are also better functionalized for transmembrane transport and intercellular interactions.

Further, one can speculate that primarily toxic plain PAMAM dendrimers may get safer toward cellular environment upon undergoing non-enzymatic N-glycosylation in the presence of surplus glucose. Indeed, researchers have demonstrated such a phenomenon in an experimental model of streptozotocin-induced diabetes mellitus (rodent model): the higher degree of hyperglycemia is more protective against the observed PAMAM toxicity [51]. Thus, it appears that the slight modifications of amino termini on PAMAM surface could be a wise option for applying them in tackling the metabolic aberrations of diabetes. Further, it looks plausible to bind lysine residues to the plain amino-terminated PAMAMs so as to obtain "polylysylated" PAMAM dendrimers (Figure 10.2). Because such dendrimers acquire two new free amino "valencies" of lysine (for each occupied primary amino group on the dendrimer surface), it may reasonably be argued that PAMAM dendrimers (modified with L-lysine) might be even more effective in the scavenging of surplus glucose compared with their unmodified counterparts.

Thus, the PAMAM dendrimers loaded with lysine residues on their surfaces could be an effective target in reducing of the post-translational modifications of critical proteins, and perhaps even more suitable for such a scavenging compared with non-lysylated dendrimers. Especially, the amino groups of terminal lysyl residues in case of the cationic "polylysylated" PAMAM dendrimers are considered as promising chemical units

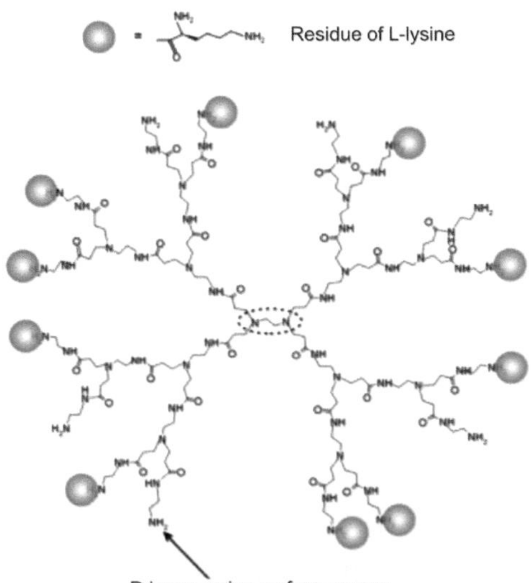

FIGURE 10.2 Poly(L-lysylated) PAMAM dendrimers. L-lysine (solid balls) can be linked to surface terminal groups either directly (via peptide bond) or via a linker. (From: Magdalena Labieniec-Watala, Cezary Watala. PAMAM Dendrimers: Destined for Success or Doomed to Fail? Plain and Modified PAMAM Dendrimers in the Context of Biomedical Applications. Journal of Pharmaceutical Sciences 104:2–14, 2015; DOI10.1002/jps.24222. http://onlinelibrary.wiley.com/doi/10.1002/jps.24222/full)

for the attachment of different functional groups. As the density of the terminal amino groups on the dendrimer surface increases (with increasing generations of PAMAM), it is expected that the attachment of other molecules to the polymer becomes rather easier [52].

The rising interest in poly-L-lysine (PLL) dendrimers has triggered a new array of investigations in the hope for their potential applications in nanobiotechnology. Earlier, a linear PLL was employed for severl years as a vector for the cellular delivery of DNA [53]. Currently, PLL dendrimers are a well-established dendrimer family [54] (Figure 10.3). Though less frequently reported than PAMAM dendrimers or the poly (propylene imine) dendrimers, PLL dendrimers have shown their potential usefulness in different biological applications including gene delivery, drug delivery, peptide antigens, antimicrobial agents and vaccines [55]. However, there is one hitch-PLL dendrimers are widely considered both cytotoxic and hemolytic. This is because they bear numerous surface lysines,

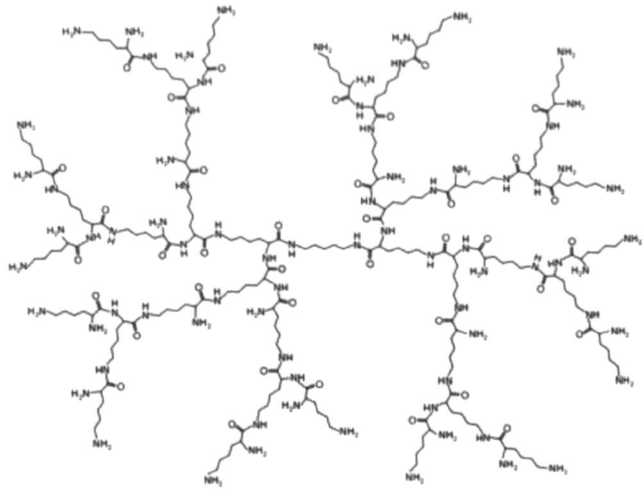

FIGURE 10.3 PLL dendrimers. (From: Magdalena Labieniec-Watala, Cezary Watala. PAMAM Dendrimers: Destined for Success or Doomed to Fail? Plain and Modified PAMAM Dendrimers in the Context of Biomedical Applications. Journal of Pharmaceutical Sciences 104:2–14, 2015; DOI10.1002/jps.24222. http://onlinelibrary.wiley.com/doi/10.1002/jps.24222/full)

i.e., they still exhibit strong polycationic characteristics. Therefore, PLL researchers often make attempts to reduce these potential adverse effects by modifying the surface of these dendrimers with nonionic groups, e.g., PEG or other chemical moieties (arylsulfonate, phenyldicarboxylate, naphthylsulfonate, fullerene, succinate, or arginine). Evidence suggest that the capping the surface of PLL dendrimers with such groups reduces the chances of vascular binding and increases *in vivo* stability, in contrast to the corresponding cationic PLL polymers [56, 57]. Moreover, it has been shown that such modifications improve the efficacy of chemotherapeutic agents in cancer treatment, when PLL dendrimers are deployed as drug carriers [58]. These polymer modifications have also been shown to effectively prolong half-life in the circulation, enhanced tumor accumulation, and protect their drug payload from possible enzymatic destruction [59]. Thus, dendrimers based on PLL residues are certainly promising candidates, next to PAMAMs, for different biomedical applications. However, more work needs to be done to rectify issues such as proteolytic instability, interaction with endothelial surfaces, blood components, or uptake by the phagocytic cells. Importantly, uncapped PLL dendrimers undergo extensive breakdown to monomeric lysine [60]. Although majority of these limitations may be circumvented by conjugation with specific chemical groups (mentioned above), the use of plain PLL dendrimers, just as in the case of plain PAMAMs, should be made with a pinch of salt.

Novel results pertaining to the toxic mechanisms of action of amine-/hydroxyl-terminated PAMAM dendrimers towards environmentally relevant microorganisms (e.g., a cyanobacterium of the genus Anabaena, and the green alga *Chlamydomonas reinhardtii*) have been reported. PAMAM ethylenediamine core dendrimers (G2 to G4) had been employed [61]. They displayed a positive charge (measurable as ζ-potential) in culture media. All the amine-terminated and especially the G4 hydroxyl-terminated dendrimer inhibited the growth of both the tested microorganisms. However, compared with the cyanobacterium, the effect on the growth of the green alga was higher. Further, there was a clear-cut relationship between dendrimer generation and its toxicity; higher toxicity was observed for higher generation. At low concentrations, hormesis was observed for hydroxyl-terminated dendrimers. The cationic dendrimers as well as G4-OH favored the formation of reactive oxygen species (ROS) in both the tested organisms, and this was not related with the chloroplast/photosynthetic membranes. Moreover, the photochemistry of Photosystem II was not affected. Cell damage resulted in cytoplasm disorganization and cellular deformities. It was demonstrated for the first time that cationic PAMAM dendrimers were quickly internalized by both microorganisms. These results warn against the generalized use of dendrimers, which may pose significant risk for the health of natural ecosystems.

A 'rate equation model' was constructed to numerically mimic nanoparticle uptake and subsequent cellular response using polyamidoamine dendrimers (generations 4–6) [62]. The temporal evolution of the intracellular cascade of increased levels of ROS, intracellular antioxidants, caspase activation, mitochondrial membrane potential decay, tumor necrosis factor and interleukin generation was simulated. The dose and generation dependence of these response factors was observed to suitably represent experimental observations at various time points. The model indicates that variations between responses of different cell-lines (such as murine macrophages, human keratinocytes and colonocytes) can be simulated. Further, within a given cell-line, varying responses of different cytotoxicity assays can be studied in terms of their sensitivities to various intracellular cascade events. The model serves as a tool to interpret the range of dose and temporal dependence besides elucidating the mechanisms underlying the cytotoxic response to nanoparticle exposure. It also describes the interaction in terms of specific nanoparticulate properties and parameters based on reaction rates. Such an approach could be useful for classification of nanotoxicity and may lay the foundation for future quantitative structure-activity relationships besides predictive nanotoxicity models.

PAMAM dendrimers are among the most extensively investigated for brain-specific drug delivery, imaging and diagnosis. Unfortunately, the neurotoxicity of PAMAM dendrimers (the underlying mechanism of which is poorly-defined) poses a great challenge to their practical applications. PAMAM dendrimers have been shown to induce both cytotoxicity and autophagic flux in human glioma cell lines [63]. Inhibition of autophagy reverses cell death indicating the cytotoxic role of autophagy in neurotoxicity. Akt/mTOR pathway probably initiates PAMAM dendrimers-induced autophagy. Moreover, autophagy may be, at least in part, mediated by intracellular ROS generation. Collectively, the data suggest the critical role of autophagy in neurotoxicity associated with cationic PAMAM dendrimers, raising valid concerns about possible neurotoxic adverse effects caused by future clinical applications of PAMAM dendrimers besides providing potential targets (Akt/mTOR signaling pathway) to ameliorate the toxic effects.

The paradigm about dendrimers' toxicity based on *in vitro* studies needs to be revised. Nearly all dendrimers of lower-middle generations are non-toxic *in vivo*, despite showing few cytotoxic effects *in vitro*. Only the higher generations of unmodified cationic dendrimers at high doses may have some toxicity *in vivo*. Some of the undesirable effects following the administration of unmodified cationic dendrimers reduce during long term dosing because of the resultant development of counteracting mechanisms so that the alterations tend to return to baseline levels during the recovery period [64]. Further, neutralization of the surface charge of dendrimers (in their dendriplexes) might lead to less toxicity *in vivo*. Chemical modifications determine the desired location of dendrimer-based conjugates/dendriplexes in a target organ. Although dendrimers/dendriplexes do accumulate temporarily in liver, pancreas, heart, and kidneys, they do not cause permanent damage, i.e., the risk of malfunction of internal organs is minimal. Clearance of dendrimers also strongly depends on their chemical nature, e.g., nucleic acids complexed with dendrimers are more stable, having longer half-life than free and PEI-complexed ones. Last but not the least, the majority of dendrimers/dendriplexes are non-immunogenic [64, 65].

10.3 NOW AND NEXT?

The foregoing facts suggest that in order to fully exploit the potential of dendrimers in medicine, an in-depth understanding of several key points is essential:

1. Despite elaborate efforts and the progress till date, the problem of a reliable method of evaluating non-toxic doses of dendrimers still remains enigmatic. Thus, an important objective for future research should be the determination of optimal doses at which dendrimers could be used with the desired efficacy.
2. Toxicity of cationic dendrimers is still a major hurdle in their preclinical and clinical applications. It has been reported time and again that dendrimers can induce irreversible damage to organs/tissues/cells in experimental models (*in vitro*, *ex vivo*, *in vivo*). Thus, the impact of dendrimers on the overall survival and on the functions of organs needs to be carefully examined.
3. Despite the fact that the number of labs conducting *in vivo* studies with dendrimers is slowly increasing, the most suitable (nontoxic as well as effective) way of dendrimer administration is yet to be developed. It goes without saying that the route of dendrimer delivery to target tissues has a great influence on the overall quality of treatment.
4. Another emerging problem is the interpretation of test results obtained in various seasons and arising from different animal strains/breeds as well as models. Perhaps a uniform and unequivocal protocol needs to be developed with an international collaborative effort.
5. Most of the conclusions concerning the properties/activity of dendrimers are drawn from *in vitro* studies. Unfortunately, the results

obtained from the *in vitro* and *in vivo* studies often are poorly correlated thereby limiting their practicability.

6. Notwithstanding the plethora of scientific reports encouraging and raising optimism for the potential biomedical use of dendrimers, their high cost of production (and consequently, the high price of the final dendrimer formulation) may pose serious constraints on the use of dendrimers in the treatment of chronic diseases.

KEYWORDS

- **autophagy**
- **dendriplexes**
- **glycosylated dendrimers**
- **hypoglycemic agents**
- ***in vivo* studies**
- **toxicity**

REFERENCES

1. Sadekar, S., & Ghandehari, H. (2012). Transepithelial transport and toxicity of PAMAM dendrimers: Implications for oral drug delivery. *Adv Drug Deliv Rev 64*, 571–588.
2. Markowicz-Piasecka, M., Luczak, E., Chalubinski, M., Broncel, M., Mikiciuk-Olasik, E., & Sikora, J. (2014). Studies towards biocompatibility of PAMAM dendrimers—Overall hemostasis potential and integrity of the human aortic endothelial barrier. *Int J Pharm 473*, 158–169.
3. Labieniec, M., & Watala, C. (2009). PAMAM dendrimers—Diverse biomedical applications. Facts and unresolved questions. *Centr Eur J Biol 4*, 434–451.
4. Thiagarajan, G., Greish, K., & Ghandehari, H. (2013). Charge affects the oral toxicity of poly(amidoamine) dendrimers. *Eur J Pharm Biopharm. 84*, 330–334.
5. Aillon, K. L., Xie, Y., El Gendy, N., Berkland, C. J., & Forrest, M. L. (2009). Effects of nanomaterial physicochemical properties on in vivo toxicity. *Adv Drug Deliv Rev 61*, 457–466.
6. Borowska, K., Wolowiec, S., Rubaj, A., Glowniak, K., Sieniawska, E., & Radej, S. (2012). Effect of polyamidoamine dendrimer G3 and G4 on skin permeation of 8-methoxypsoralene—In vivo study. *Int J Pharm 426*, 280–283.

7. Gupta, U., Agashe, H. B., Asthana, A., & Jain, N. K. (2006). A review of in vitro–in vivo investigations on dendrimers: The novel nanoscopic drug carriers. *Nanomedicine 2*, 66–73.
8. Malik, N., Wiwattanapatapee, R., Klopsch, R., Lorenz, K., Frey, H., Weener, J. W., Meijer, E. W., Paulus, W., & Duncan, R. (2000). Dendrimers: Relationship between structure and biocompatibility in vitro, and preliminary studies on the biodistribution of 125I-labeled polyamidoamine dendrimers in vivo. *J Control Release 65*, 133–148.
9. Roberts, J. C., Bhalgat, M. K., & Zera, R. T. (1996). Preliminary biological evaluation of polyamidoamine (PAMAM) Starburst dendrimers. *J Biomed Mater Res 30,* 53–65.
10. Labieniec-Watala, M., & Watala, C. (2015). PAMAM dendrimers: Destined for success or doomed to fail? Plain and modified PAMAM dendrimers in the context of biomedical applications. *J Pharm Sci 104*, 1–14.
11. Yao, H. R., Veine, D. M., Zeng, Z. Z., Fay, K. S., Staszewski, E. D., & Livant, D. L. (2010). Increased potency of the PHSCN dendrimer as an inhibitor of human prostate cancer cell invasion, extravasation, and lung colony formation. *Clin Exp Metastasis 27*, 173–184.
12. Chen, H. T., Neerman, M. F., Parrish, A. R., & Simanek, E. E. (2004). Cytotoxicity, hemolysis, and acute in vivo toxicity of dendrimers based on melamine, candidate vehicles for drug delivery. J Am Chem Soc 126, 10044–10048.
13. Padilla De Jesus, O. L., Ihre, H. R., Gagne, L., Frechet, J. M., Jr., & Szoka, F. C. (2002). Polyester dendritic systems for drug delivery applications: In vitro and in vivo evaluation. *Bioconjug Chem 13*, 453–461.
14. Labieniec-Watala, M., & Siewiera, K. (2013). The impact of seasonal fluctuations on rat liver mitochondria response to tested compounds—A comparison between autumn and spring. new insight into collecting and interpretation of experimental data originating from different seasons. *CellBio 2*, 20–30.
15. Labieniec-Watala, M., Siewiera, K., & Jozwiak, Z. (2011). Resorcylidene aminoguanidine (RAG) improves cardiac mitochondrial bioenergetics impaired by hyperglycaemia in a model of experimental diabetes. *Int J Mol Sci 12*, 8013–8026.
16. Siewiera, K., & Labieniec-Watala, M. (2012). Ambiguous effect of dendrimer PAMAM G3 on rat heart respiration in a model of experimental diabetes—Objective causes of laboratory misfortune or unpredictable G3 activity? *Int J Pharm 430,* 258–265.
17. Diaz, E., Vazquez, N., Fernandez, C., Durand, D., Lasaga, M., Debeljuk, L., & Diaz, B. (2011). Seasonal variations of substance P in the striatum of the female rat are affected by maternal and offspring pinealectomy. *Neurosci Lett 492*, 71–75.
18. Konior, A., Klemenska, E., Brudek, M., Podolecka, E., Czarnowska, E., & Beresewicz, A. (2011). Seasonal superoxide overproduction and endothelial activation in guinea-pig heart; seasonal oxidative stress in rats and humans. *J Mol Cell Cardiol 50,* 686–694.
19. Vazquez, N., Diaz, E., Fernandez, C., Jimenez, V., Esquifino, A., & Diaz, B. (2007). Seasonal variations of gonadotropins and prolactin in the laboratory rat. Role of maternal pineal gland. *Physiol Res 56*, 79–88.
20. Labieniec, M., Ulicna, O., Vancova, O., Kucharska, J., Gabryelak, T., & Watala, C. (2009). Effect of poly (amido) amine (PAMAM) G4 dendrimer on heart and liver mitochondria in an animal model of diabetes. *Cell Biol Int 34*, 89–97.

21. Graham, M. L., Janecek, J. L., Kittredge, J. A., Hering, B. J., & Schuurman, H. J. (2011). The streptozotocin-induced diabetic nude mouse model: Differences between animals from different sources. *Comp Med 61,* 356–360.
22. Jevprasesphant, R., Penny, J., Jalal, R., Attwood, D., McKeown, N. B., & D'Emanuele, A. (2003). The influence of surface modification on the cytotoxicity of PAMAM dendrimers. *Int J Pharm 252*, 263–266.
23. Pan, G. F., Lemmouchi, Y., Akala, E. O., & Bakare, O. (2005). Studies on PEGylated and drug-loaded PAMAM dendrimers. *J Bioact Compat Polym 20,* 113–128.
24. Kim, Y., Klutz, A. M., & Jacobson, K. A. (2008). Systematic investigation of polyamidoamine dendrimers surface-modified with poly(ethylene glycol) for drug delivery applications: Synthesis, characterization, and evaluation of cytotoxicity. *Bioconjug Chem 19,* 1660–1672.
25. Yang, H., Lopina, S. T., DiPersio, L. P., & Schmidt, S. P. (2008). Stealth dendrimers for drug delivery: Correlation between PEGylation, cytocompatibility, and drug payload. *J Mater Sci Mater Med 19*, 1991–1997.
26. Harris, J. M., Chess, R. B. (2003). Effect of pegylation on pharmaceuticals. *Nat Rev Drug Discov 2*, 214–221&
27. Wang, W., Xiong, W., Wan, J. L., Sun, X. H., Xu, H. B., & Yang, X. L. (2009). The decrease of PAMAM dendrimer-induced cytotoxicity by PEGylation via attenuation of oxidative stress. *Nanotechnology 20,* 105103.
28. Kolhatkar, R. B., Kitchens, K. M., Swaan, P. W., & Ghandehari, H. (2007). Surface acetylation of polyamidoamine (PAMAM) dendrimers decreases cytotoxicity while maintaining membrane permeability. *Bioconjug Chem 18*, 2054–2060.
29. Waite, C. L., Sparks, S. M., Uhrich, K. E., & Roth, C. M. (2009). Acetylation of PAMAM dendrimers for cellular delivery of siRNA. *BMC Biotechnol 9*, 38.
30. Luo, D., Haverstick, K., Belcheva, N., Han, E., Saltzman, W. M. (2002). Poly (ethylene glycol)-conjugated PAMAM dendrimer for biocompatible, high-efficiency DNA delivery. *Macromolecules 35*, 3456–3462.
31. Ciolkowski, M., Petersen, J. F., Ficker, M., Janaszewska, A., Christensen, J. B., Klajnert, B., & Bryszewska, M. (2012). Surface modification of PAMAM dendrimer improves its biocompatibility. *Nanomedicine 8*, 815–817.
32. Mahajan, P., Sodhi, K. S., Pandey, R., & Singh, J. (2014). Glycotherapeutics: Clinical implications. *Int J Adv Pharm Biol Chem 3*(3), 786–795.
33. Voit, B., & Appelhans, D. (2010). Glycopolymers of various architectures—More than mimicking nature. *Macromol Chem Phys 211*, 727–735.
34. Roy, R., Zanini, S., Meunier, J., & Romanowska, A. (1993). Solid phase synthesis of dendritic sialoside inhibitors of influenza A virus haemagglutinin. *J Chem Soc Chem Commun 24*, 1869–1872.
35. Chabre, Y. M., & Roy, R. (2008). Recent trends in glycodendrimer syntheses and applications. *Curr Top Med Chem 8*, 1237–1285.
36. Chabre, Y. M., & Roy, R. (2010). Design and creativity in synthesis of multivalent neoglycoconjugates. *Adv Carbohydr Chem Biochem 63,* 165–393.
37. Imberty, A., Chabre, Y. M., & Roy, R. (2008). Glycomimetics and glycodendrimers as high affinity microbial anti-adhesins. *Chemistry 14*, 7490–7499.
38. Okada, M., & Aoi, K. (1995). Synthesis of sugar-containing macromonomers by living ring-opening polymerization. *J Macromol Sci Pure Appl Chem A32*, 907–914.

39. Aoi, K., Itoh, K., & Okada, M. (1995). Globular carbohydrate macromolecule sugar balls.1. Synthesis of novel sugar-persubstituted poly (amido amine) dendrimers. *Macromolecules 28,* 5391–5393.
40. Dubber, M., & Lindhorst, T. K. (2001). Trehalose-based octopus glycosides for the synthesis of carbohydrate-centered PAMAM dendrimers and thiourea-bridged glycoclusters. *Org Lett 3,* 4019–4022.
41. Vargas-Berenguel, A., Ortega-Caballero, F., Santoyo-Gonzalez, F., Garcia-Lopez, J. J., Gimenez-Martinez, J. J., Garcia-Fuentes, L., & Ortiz-Salmeron, E. (2002). Dendritic galactosides based on a beta-cyclodextrin core for the construction of site-specific molecular delivery systems: Synthesis and molecular recognition studies. *Chemistry 8,* 812–827.
42. Svenson, S. (2015). The dendrimer paradox – high medical expectations but poor clinical translation. *Chem Soc Rev. 44*(12), 4131–44.
43. Turnbull, W. B., & Stoddart, J. F. (2002). Design and synthesis of glycodendrimers. *J Biotechnol 90,* 231–255
44. Oliveira, J. M., Kotobuki, N., Marques, A. P., Pirraco, R. P., Benesch, J., Hirose, M., Costa, S. A., Mano, J. F., Ohgushi, H., & Reis, R. L. (2008). Surface engineered carboxymethylchitosan/poly (amidoamine) dendrimer nanoparticles for intracellular targeting. *Adv Funct Mater 18,* 1840–1853.
45. Liu, K. C., & Yeo, Y. (2013). Zwitterionic chitosan-polyamidoamine dendrimer complex nanoparticles as a pH-sensitive drug carrier. *Mol Pharm 10,* 1695–1704.
46. Salgado, A. J., Oliveira, J. M., Pirraco, R. P., Pereira, V. H., Fraga, J. S., Marques, A. P., Neves, N. M., Mano, J. F., Reis, R. L., & Sousa, N. (2010). Carboxymethylchitosan/poly (amidoamine) dendrimer nanoparticles in central nervous systems-regenerative medicine: Effects on neuron/glial cell viability and internalization efficiency. *Macromol Biosci 10,* 1130–1140.
47. Princz, M. A., & Sheardown, H. (2012). Heparin-modified dendrimer cross-linked collagen matrices for the delivery of heparin-binding epidermal growth factor. *J Biomed Mater Res A 100,* 1929–1937.
48. Woller, E. K., & Cloninger, M. J. (2001). Mannose functionalization of a sixth generation dendrimer. *Biomacromolecules 2,* 1052–1054.
49. Ashton, A. R., Boyd, S. E., Brown, C. L., Nepogodiev, S. A., Meijer, E. W., Eerlings, H. W. I., & Stoddart, J. F. (1997). Synthesis of glycodendrimers by modification of poly(propylene imine) dendrimers. *Chem Eur J 3,* 974–984.
50. Barata, T. S., Shaunak, S., Teo, I., Zloh, M., & Brocchini, S. (2011). Structural studies of biologically active Glycosylated polyamidoamine (PAMAM) dendrimers. *J Mol Model 17,* 2051–2060.
51. Labieniec, M., Ulicna, O., Vancova, O., Glowacki, R., Sebekova, K., Bald, E., Gabryelak, T., & Watala, C. (2008). PAMAM G4 dendrimers lower high glucose but do not improve reduced survival in diabetic rats. *Int J Pharm 364,* 142–149.
52. Demirci, S., Emre, F. B., Ekiz, F., Oguzkaya, F., Timur, S., Tanyeli, C., & Toppare, L. (2012). Functionalization of poly-SNS-anchored carboxylic acid with Lys and PAMAM: Surface modifications for biomolecule immobilization/stabilization and bio-sensing applications. *Analyst 137,* 4254–4261.
53. Kataoka, T., Yamamoto, S., Yamamoto, T., & Tokunaga, T. (1990). Immunotherapeutic potential in guinea-pig tumor model of deoxyribonucleic acid from mycobacterium

bovis BCG complexed with poly-L-lysine and carboxymethylcellulose. *Jpn J Med Sci Biol 43*, 171–182.
54. Roberts, B. P., Scanlon, M. J., Krippner, G. Y., & Chalmers, D. K. (2009). Molecular dynamics of poly (L-lysine) dendrimers with naphthalene disulfonate caps. *Macromolecules 42,* 2775–2783.
55. Ohsaki, M., Okuda, T., Wada, A., Hirayama, T., Niidome, T., & Aoyagi, H. (2002). In vitro gene transfection using dendritic poly (L-lysine). *Bioconjugate Chem 13*, 510–517.
56. Kaminskas, L. M., Boyd, B. J., Karellas, P., Henderson, S. A., Giannis, M. P., Krippner, G. Y., & Porter, C. J. H. (2007). Impact of surface derivatization of poly-L-lysine dendrimers with anionic aryisulfonate or succinate groups on intravenous pharmacokinetics and disposition. *Mol Pharm 4,* 949–961.
57. Kaminskas, L. M., Boyd, B. J., Karellas, P., Krippner, G. Y., Lessene, R., Kelly, B., & Porter, C. J. H. (2008). The impact of molecular weight and PEG chain length on the systemic pharmacokinetics of PEGylated poly L-lysine dendrimers. *Mol Pharm 5,* 449–463.
58. Al Jamal, K. T., Al Jamal, W. T., Wang, J. T., Rubio, N., Buddle, J., Gathercole, D., Zloh, M., & Kostarelos, K. (2013). Cationic poly-L-lysine dendrimer complexes doxorubicin and delays tumor growth in vitro and in vivo. *ACS Nano 7*, 1905–1917.
59. Fox, M. E., Guillaudeu, S., Frechet, J. M. J., Jerger, K., Macaraeg, N., & Szoka, F. C. (2009). Synthesis and in vivo antitumor efficacy of pegylated poly(L-lysine) dendrimer-camptothecin conjugates. *Mol Pharm 6*, 1562–1572.
60. Kaminskas, L. M., Wu, Z., Barlow, N., Krippner, G. Y., Boyd, B. J., & Porter, C. J. (2009). Partly-PEGylated poly-L-lysine dendrimers have reduced plasma stability and circulation times compared with fully PEGylated dendrimers. *J Pharm Sci 98,* 3871–3875.
61. Gonzalo, S., Rodea-Palomares, I., Leganés, F., García-Calvo, E., Rosal, R., & Fernández-Piñas, F. (2014). First evidences of PAMAM dendrimer internalization in microorganisms of environmental relevance: A linkage with toxicity and oxidative stress. Nanotoxicology. Oct 17, 1–13. [Epub ahead of print].
62. Maher, M. A., Naha, P. C., Mukherjee, S. P., Byrne, H. J. (2014). Numerical simulations of in vitro nanoparticle toxicity – the case of poly (amido amine) dendrimers. *Toxicol in Vitro. 28*(8), 1449–1460.
63. Wang, S., Li, Y., Fan, J., Wang, Z., Zeng, X., Sun, Y., & Song, P., Ju, D. (2014). The role of autophagy in the neurotoxicity of cationic PAMAM dendrimers. *Biomaterials. 35*(26), 7588–7597.
64. Shcharbin, D., Janaszewska, A., Klajnert-Maculewicz, B., Ziemba, B., Dzmitruk, V., Halets, I., Loznikova, S., Shcharbina, N., Milowska, K., Ionov, M., Shakhbazau, A., & Bryszewska, M. (2014). How to study dendrimers and dendriplexes, I. I. I., Biodistribution, pharmacokinetics and toxicity in vivo. *J Control Release. 181*, 40–52.
65. Liu, Y., Tee, J. K., & Chiu, G. N. (2015). Dendrimers in oral drug delivery application: current explorations, toxicity issues and strategies for improvement. *Curr Pharm Des. 21*(19), 2629–2642.

CHAPTER 11

EMERGING ASPECTS IN DENDRIMER RESEARCH

ZAHOOR AHMAD PARRY, PhD, and RAJESH PANDEY, MD

CONTENTS

11.1 INTRODUCTION

All sciences are dynamic in nature, i.e., new concepts and applications emerge with time and this has been seen in more recent times especially in applications in medical field owing to their direct impact on human health. Like many other sciences with medical applications the science of dendrimers is fast evolving and a number of potential research teams are working globally so that newer and newer knowledge is added to this field. More and more research is focusing on all aspects of dendrimer research including structure, affinity and applications as under:

1. It is generally believed that terminal groups of dendrimers have a crucial role in determining the properties of dendrimers and the internal structure is considered to be more or less inert with less functional role. Recent studies have shown that this assertion is highly misleading, using biological data generated from human monocyte activation and all-atom molecular dynamic simulations on different family of dendrimers (having identical terminal groups but different internal structures), research has demonstrated that the inner core scaffold strongly influences overall dendrimer properties.
2. There are several medical conditions were medicines have limited success owing to pharmacokinetic problems mainly related to residence time and organ specific delivery. In this regard, recently, dendrimers have been exploited by conjugating the active principle (drug/antibody, etc.) with them to produce new entities called Dendribodies. Such a principle has been applied form ethamphetamine (METH) abuse in which anti-METH single chain antibody (scFv7F9Cys) was conjugated to dendrimer nanoparticles via a polyethyleneglycol (PEG) linker. High affinity and specificity of antibody was unchanged after nanoparticle conjugation on one hand and drastic improvements in serum levels and residence time were observed on the other hand in rats along with the desired organ distribution. This study laid the foundation stone for dendribody design for generating multivalent antibodies with customizable pharmacokinetic profiles against other medical conditions.
3. An unparallel challenge in the field of supra molecular chemistry is questionable affinity of polar molecules in aqueous solutions. The major issue is how supra molecular principles can be used to drive and control carbohydrate recognition. During recent times, dendritic side chains have been shown to moderate the binding properties of synthetic carbohydrate receptors. In this direction, a unique example has been developed using dendrimers for deliberate lypositioning polar groups for improved carbohydrate binding in aqueous systems. This strategy holds immense potential in various field of chemistry.
4. Worldwide, one of the worst dental conditions is dentin hypersensitivity that gives unbearable pain owing to dentin exposure to

external stimuli. Occlusion of dentinal tubules by means of mineralization is an effective method to alleviate this problem. A number of chemical agents that containing fluoride, strontium salts, oxalate, glutaraldehyde and bioactive glass are used for this purpose but limitations like short durability, poor effectiveness and incomplete occlusion effects limit their use. Dendrimers have been shown to infiltrate dentinal tubules and induce hydroxyapatite formation that leads to effective tubule occlusion. Additionally, dendrimers were shown to induce remineralization of demineralized dentin and thus have the potential in dentinal tubule occlusion. Clearly, this in vitro study created a new hope of a novel therapeutic technique for the treatment of dentin hypersensitivity.

5. Who is not aware of the havoc created by emergence and worldwide spread of drug resistant pathogens? This has attracted researchers globally for developing new drugs and as such a range of molecules is successfully in place towards clinics. One such class of compounds is antimicrobial peptides of natural or synthetic origin. However, the major challenge for this class of compounds is in vivo stability, half-life and antimicrobial potential. Recently, peptide based dendrimers have been developed that exhibit higher antimicrobial potency than natural counterparts and simultaneously can escape pathogenic defense systems. This research is going to unfold a new chapter in antibiotic development and has to play a pivotal role for human health.

In the subsequent section, the above new concepts regarding dendrimers have been discussed in detail under following headings:

1. Dendrimer scaffold;
2. Dendribodies;
3. Dendrimer-based affinity enhancement;
4. Dentistry;
5. Drug resistance.

11.2 DENDRIMER SCAFFOLD

Owing to potential applications of dendrimers due to their terrific biological properties, a large number of studies are being carried out on dendrimers each year [1]. Most of these studies focus on the number of terminal groups and/or on their modification for multivalency effect which in turn is particularly important for biological applications and thus considered to be the most important property of dendrimers [2–6]. Since three of five proposed critical nanoscale design parameters (size, shape, surface chemistry, flexibility and architecture) are related to the internal structure, therefore it is surprising to see that only few publications have experimentally focused till date on the influence of the internal structure of dendrimers on their properties. Difference in branch length among dendrimers has been considered to be the most important characteristic for obtaining nanoparticles while examining different dendrimers [7–9]. Important differences have been reported between PAMAM and poly (L-lysine) dendrimers after careful physical examination [10]. Rigidity of branches especially in dendrimers with azobenzene core has been shown to induce significant differences for the isomerization [11].

Biologically, using transfection experiments, specific dendrimers have been compared with PAMAM dendrimers [12]. Poly (phosphorhydrazone) based dendrimers with azabis (phosphonic acid) end groups have been shown to exhibit unprecedented biological properties. These dendrimers are able to activate monocytes via an anti-inflammatory pathway and hence induce multiplication of natural killer cells under *in vitro* conditions and thus can effect human immune response [13–16]. Indeed this molecule proved its efficacy in a murine model of human rheumatoid arthritis [17]. It has been shown that azabis (phosphonic acid) ended dendrimer targeting of monocytes/macrophages leads to inhibition of physiopathology of the disease that includes systemic inflammation, cartilage degradation and bone resorption.

Interestingly, initial reports on this dendrimer showing replacement of phosphonic acid end groups with carboxylic acids or sulfonic acids did not affect its activity thereby ruling out the role of end groups. On the contrary, it has been shown that within its structure, the azabisphosphonic (N(CH2P(O)(OH)(ONa)) pincer is the most active part; variation on the structure of which significantly decreases the biological activity [13, 18]. Further research showed that azabisphosphonic pincer based dendrimers are not drug carriers but drugs by themselves owing to the fact that lowering their 12 terminal groups to 8 or 10 does not affect their activity but they become poorly active upon further decrease and indeed the monomer is not active at all. It is also interesting to note that increasing number of terminal functions beyond 16 proves detrimental for activity [13, 19, 20].

These studies provided a platform for investigating the influence of scaffold nature on the properties of dendritic nanodrugs using above terminal groups. Indeed such studies were followed by comparing activation potential of seven different families of dendrimers exhibiting varied internal structures [PAMAM, PPI, poly(carbosilane), poly(L-Lysine)and three different types of phosphorus-containing dendrimers] but identical terminal groups [4–6] and the functions of azabisphosphonic acids with respect to human monocytes [21]

These dendrimers activated monocytes to different extents based on monocyte size and granularity increase thereby clearly ruling out any significant role of number of terminal functions. Rather, the rather internal structure seems to be very critical for activation potential. These studies further revealed that above dendrimers on the basis of chemical structure could be divided into two families: (i) having aromatic groups in their structure; and (ii) those having essential alkyl linkages. All aromatics dendrimers have been found to be active and various dendrimers bearing alkyl linkages also exhibited activity.

To explain these puzzling results, all-atom MD simulations have been employed for the above dendrimers immersed in a solvation box containing explicit water molecules and salt ions to gain their molecular-level description in the real environment. These simulation studies demonstrated that size like terminal groups has non-significant role in controlling molecular activity. It has been hypothesized based on simulation results that dendrimer activity is related to overall shape or configuration of dendrimers in solution owing to the fact that the structure of the active dendrimers is segregated in salt water with the entire hydrophilic terminal functions compacted on one side and the hydrophobic scaffold exposed to the external media. On the other hand, non-active dendrimers exhibit symmetrical structure with terminal groups displayed all around the dendrimer surface. Accordingly, dendrimers have been divided into two categories, directional and spherical based on configuration. Keeping in view that there are different factors in many different diseases the biological potential of several dendrimers with respect to human monocyte activation has been studied for better quantifying structure–activity relationships of dendrimers using flow cytometry [13–17]. These studies showed that varying linker length has no effect within a particular dendrimer series. Simulation studies backed by biological ones suggest that both hydrophilicity/hydrophobicity and molecular directionality are the most critical parameters for biological activity. Indeed, the best active dendrimers possess the most favorable solvation energy and the highest directionality scores, thereby reflecting direct correlation between the dendrimers' multivalency and activity. Bio-molecules like proteins follow the same concept in which surface features like presence of charged or active patches, local and overall hydrophobicity and hydrophilicity results in recognizable effects via innumerable interactions and molecular recognition. Clearly, these studies show that changing a single parameter within the internal structure of dendrimers can totally modify its properties.

11.3 DENDRIBODIES

In order to make treatment outcomes significantly more effective recent years have witnessed the use of nanotechnology in a new way in which the active principle, e.g., an antibody is conjugated to dendrimers via linkers to generate high-order conjugates termed Dendribodies. Such Dendribodies have been prepared for the treatment of methamphetamine (METH) abuse by conjugating an anti-METH single chain antibody to dendrimers via polyethyleneglycol. In this strategy, single chain variable fragments (scFv) of anti-METH antibody were conjugated with polyamidoamine (PAMAM) dendrimer nanoparticles using polyethyleneglycol (PEG) as linker to generate multivalent G3-PEG-scFv [22]. Further studies have shown that pharmacokinetic and efficacy profile of an anti-METH antibody scFv7F9Cys was significantly better in conjugated dendribody compared to unconjugated one. This prolonged activity of conjugated scFv7F9Cys dendrimer *in vivo* was found to be due to increased residence time owing to its lower clearance rate compared to unconjugated counterpart. One of the critical questions in this area was how to increase $t_{1/2}$ of these therapeutic conjugates especially with rise of antibody fragments. Several strategies were developed using two major approaches to enhance overall pharmacokinetic profiles of scFvs [23, 24], first one being multimerization using recombinant manipulation but high self association potential of scFvs-led to formation of different molecular weight complexes which in turn leads to production problems and hence variable therapeutic properties [23, 24]. Second one is chemical conjugation using PEG and this strategy has proven improved *in vivo* half-life, however this strategy neither increased scFv potency nor increase its binding valency [25, 26]. Further, PEG conjugation has been shown to decrease the affinity of some conjugated scFv [27, 28].

Recent successful attempts have been made to design adendribody exploiting a single METH binding scFv for multivalent Nanomedicine with the capacity to bind multiple METH molecules along with decent pharmacokinetic profiles of experimental medication [29, 30]. Functionality and serum pharmacokinetic profiles of this dendribody were tested *in vivo* using rat model of chronic METH. Serum concentration versus time profiles of un-conjugated scFv7F9Cys as well as of dendribody based were biphasic curves that fit a two-compartment pharmacokinetic model. Pharmacokinetic results revealed that Cls and volume of distribution (Vd) of scFv7F9Cys in dendribody form decreased 43 and 1.6 fold respectively compared to that of un-conjugated form but terminal elimination half-life ($t1/2\lambda z$) increased 20 fold. This terrific $t_{1/2}$ improvement is clearly owing to reduced Cls of scFv7F9Cys in dendribody form. Noteworthy from this study was also the finding that both conjugated and un-conjugated forms of scFv7F9Cys were restricted to serum only and had no association with erythrocytes. Therefore, unconjugated scFv7F9Cys and in dendribody form had no evident toxic effects, this is further backed by *in vitro* hemolysis data and the safety studies of the PEG modified dendrimers. With unconjugated scFvs, some accumulation in the right kidney at 0.5 hr upon 7% of the injected dose per gram of tissue was observed probably owing to re-absorption by the kidney tubules as it was not found later possibly due to metabolism after re-absorption [31]. In contrast to dendribody group, only some splenic accumulation was observed at early time points at injected dose of 0.6%; no other significant difference was observed in two groups.

Therefore, crystal clear *in vivo* evidence exists in support of improved pharmacokinetics and efficacy of any active principle like anti-METH scFv by dendribody approach; this can be exploited against a number of diseases in which active principle requires improved pharmacokinetics and organ-specific targeting. Another recent most approach has been exploitation of Dendribodies for critical diagnosis where traditional approaches had only limited success. An extraordinary example of such an approach for colon cancer is discussed below in detail.

Cancer relapse and metastasis owing to circulating tumor cells (CTCs) is a leading cause of world wide cancer-related deaths [32–34]. CTCs are cells in the blood stream of early-stage cancer patients that originate from primary tumors or metastatic sites [34–37]. These cells are usually at extreme low concentrations with respect to non-cancerous hematopoietic cells (1 in 10^6 to 10^9 cells) in cancer survivors at remission stage [35, 38, 39] and they lack proliferation and invasion capacity. But CTCs get activated in specific microenvironment and turn into disseminated tumor cells and metastasis-initiating cells and that in turn mediate hematogenous spread of cancer to distant sites and progression of cancer metastasis respectively [35, 40–42].

CTCs bear specific information related to primary tumors and express biomarkers which are not expressed by normal and carcinoma cells [36, 43, 44]. Therefore, change in CTCs numbers in blood directly reflects metastatic progression or survival potential of patient [43, 45, 46]. Keeping in view the importance of CTCs as potential markers of patient prognosis, researchers have exploited several approaches to isolate and quantify CTCs from other large interfering cell populations [47–59] (Table 11.1).

Exploiting unique properties of dendrimers including multivalent conjugation potential and capacity to interact specifically (antigen and antibody), studies were designed to prepare anti-Slex-coated dendrimer conjugates for capturing and quantifying CTCs *in vitro*. Interestingly, CTCs were captured in artificial blood with high efficiency on one hand and exhibited compromised cell activity in dendrimer conjugated forms on the other hand. The latter observation can be used as a new platform for effective prevention of metastasis. For instance, a well designed procedure was developed using multiple aSlex antibodies for assembling nanosized PAMAM dendrimers via multivalent binding capacities; aSlex-conjugated dendrimers were synthesized as a single entity [60, 61]. High-affinity molecular recognition FITC-labeled G6-aSlex conjugates

TABLE 11.1 Approaches and Drawbacks Related to Isolation of CTCs

Approaches
• Immunomagnetic separation
• Microfluidics-based technologies
• Filtration
• Affinity-based surface capture
Drawbacks
• Low capture yield
• Time-consuming
• Non-specific binding
• Tight binding
• Miss smaller CTCs

produced had binding ability in biological processes. It was shown that optimum capture efficiency was achieved 1 h after HT29 cells were incubated with G6–5aSlex-FITC conjugate at 20 mg/mL. More specific binding and capturing of adherent and non-adherent HT29 cells was observed in G6–5aSlex-FITC conjugate within 1-h compared to that of controls. G6–5aSlex-FITC conjugate showed the higher capture efficiency with or without pretreatment of aSlex-FITC than FITC-G6- COOH-5aSlex conjugate possibly owing to availability of carboxyl groups. Further capture efficiency of G6–5aSlex-FITC conjugate was decreased by significantly increasing concentrations of interfering cells (HL-60 and RBCs) possibly due to huge blockage from interfering cells and the reduced interaction chance between conjugate and the target cells [62]. Analysis of cell anti-proliferation, cell cycle stagnation, cell apoptosis demonstrated that aSlex-coated G6 PAMAM dendrimer conjugates exhibit regulation properties that restrain cell activity without killing cells. Overall results collectively demonstrated that aSlex conjugated-dendrimers could specifically recognize and capture the Slex-expressing HT29 cells not RBCs and multiple conjugation of aSlex to dendrimer surfaces largely contributed to the improved capture efficiency and the enhanced down-regulation of cell activity in comparison with the effect of monovalent conjugates. This study laid the new foundation stone with respect to diagnostic and therapeutic potential of dendrimers and undoubtedly more literature would emerge in this direction with time.

11.4 DENDRIMER-BASED AFFINITY ENHANCEMENT

One of the critical challenges in the field of supra molecular chemistry is the binding of polar molecules in aqueous solution. The major focus of this research is to explore in both natural and synthetic systems how supra molecular principles and particular interactions may be used to drive and control carbohydrate recognition [63]. At present it is not clear how polar binding groups can be used to increase affinities in water because such studies are limited owing to design problems and related synthetic queries. Basic problem comes owing to the fact that altering amide groups that are responsible for polar interactions would change hydrophobic surfaces and, on the contrary, introduction of polar groups at receptors requires difficult synthetic efforts. Recognizing this problem, an expert opinion was that changes can be made in solubilizing groups because such modifications are expected to make little difference with respect to binding properties. Indeed this concept received experimental evidence in which side chains were solubilized in synthetic carbohydrate receptors and binding enhanced. This process was exploited to produce highest affinities reported till date for biomimetic carbohydrate recognition in water [64].

A second-generation dendrimer constructed from tert-butylacrylate and tris-(hydroxymethyl) amino methane was used to investigate glucose binding in water by ^{1}HNMR spectroscopy. This investigation revealed connections between side-chain hydrogen atoms and inward-directed receptor hydrogen atoms there by implying that terminal strands from the side chain can thread through the cavity. Modeling further confirmed that such threading was indeed possible. These results proved that dendrimeric side chains can influence binding properties. This principle was used to synthesize a range of receptors with controlled side-chain length, charge, and steric bulk. The results of those studies revealed that tuning of side chains in receptors can indeed be done to enhance binding properties and adjust selectivity.

Hence, these recent efforts confirm that dendritic side chains could be exploited for enhancing binding properties of synthetic carbohydrate receptors and can also serve as solubilizing groups. This strategy has been successfully used to increase affinities for glucosamine. Therefore, a unique strategy based on the use of dendrimers for deliberate positioning of polar groups to enhance carbohydrate binding in water has been developed. Dendritic side chains are expected to play crucial role in future designs that ensure water solubility for hydrophobic core structures.

11.5 DENTISTRY

One of the most predominant dental conditions in clinic is dentin hypersensitivity characterized by sharp pain owing to dentin exposure to external stimuli [65, 66]. Dentin hypersensitivity is generally believed to be due to a pain-provoking stimulus owing to changes in dentinal tubular fluid (increase or direction change) that in turn stimulates the nerves around the odontoblasts [67]. This hypersensitivity is commonly treated either by occluding the dentinal tubules to reduce the fluid flow or by decreasing the excitability of the intradental nerve using potassium ions containing chemicals [68, 69]. Different chemical agents that containing fluoride, strontium salts, oxalate, glutaraldehyde and bioactive glass are used to treat dentin hypersensitivity [70–74]. Though these agents are effective but short durability, poor effectiveness and incomplete occlusion effects limit their use [75, 76]. Therefore, search for new agents or newer methods that can solve these problems have already started.

PAMAM dendrimers exhibiting unique and well-defined secondary structures have the potential to act as mimics of proteins and thus considered to be good candidates for studying in organic crystallization. This is especially true for calcium phosphate and hydroxyapatite as the morphology and size of hydroxyapatite has been shown to be regulated under hydrothermal conditions by terminal functional groups of PAMAM dendrimers [77, 78]. PAMAM dendrimers owing to layered structures can be easily modified with respect to desired targets [79, 80]. PAMAM dendrimers have therefore been used to induce dentinal tubule occlusion (HAP formation) by two different experimental approaches. Human third molars (caries-free) were extracted, collected, cleaned and stored in 0.5% thymol at 4°C up to one month before use [81]. 1 mm dentin discs were prepared by two parallel cuts perpendicular to the long axis of each tooth just above the cement-enamel junction (CEJ). Discs were carefully examined to ensure that they were exposure free. Treatment was preceded in following four groups: (1) treated with 10% citric acid, (2) first treated with citric acid followed by coating with pure G3.0 PAMAM dendrimer, (3) demineralization using 0.5 MEDTA solution under shaking conditions at room temperature for 72 hours, (4) EDTA demineralization was followed with PAMAM- glutaraldehyde (0.25% glutaraldehyde containing 0.5% G3.0 PAMAM dendrimer) treatment at 4°C for 24. In each group, samples were rinsed with deionized water at the end. Simulated body fluid (SBF) with pH 7.25 at 36.5°C, bearing main components and ionic strength close to that of human saliva (2.5 mmol/LCa^{2+} and 1.0 mmol/LHPO_4^{2-}) were used inducing mineralization. Each sample was immersed in SBF under shaking conditions at 37°C for 7 days and SBF was replaced after every two days. PAMAM dendrimers hold promising potential in occluding dentinal tubules with *in situ* formed HAP owing to presence of higher number of amino groups in them that are able to induce mineralization in SBF [82, 83]. PAMAM dendrimers indeed were observed to be either trapped on dentin surface or infiltrated dentinal tubules and attracted PO_4, an important feature for mineralization. PAMAM dendrimer induced precipitate efficiently occluded dentinal tubules at both the exterior open end and in the depth of dentinal tubules. Such excellent results of adherence between dentin and PAMAM dendrimers are due to flexibility of dendrimers, localization of PAMAM dendrimers to dentin surface owing to amine terminated groups and potential of PAMAM dendrimers owing to positive charge to interact with negatively charged sites of hydroxyapatite crystals. Liquid nature of PAMAM dendrimer induces nucleation and growth of mineral crystals available for occlusion by infiltrating deeper in dentinal tubules. There were two drawbacks in this study: (1) occlusion of dentinal tubules using PAMAM dendrimers is a relatively slow process that would not be accepted by patient suffering from in tolerable discomfort; and (2) cross linking agent used glutaraldehyde has some cytotoxicity. These issues can be addressed by including potassium ions and by replacing glutaraldehyde with other cross-linking agents with known biocompatibility. This unique biocompatibility coupled with minimal toxicity could make PAMAM dendrimers to hold significant potential for future application in patients suffering from miseries related to their oral cavities [84, 85].

11.6 DRUG RESISTANCE

One of the greatest worldwide public health threats is the development and spread of drug resistant microbes. Therefore, search for new antibiotics to eradicate resistant microbial strains has become the utmost priority globally. Innumerable molecules from every class of compounds of natural or synthetic origin have been exploited for this purpose, one such class being amphiphilic antimicrobial peptides. These act in a non-specific manner and represent a great source of inspiration [86]. Extensive research has established the fact that these peptides act by binding and destabilizing cellular membranes [87–91]. In recent years, an interesting way to curb antibiotic resistance has been developed by synthesizing peptide-based dendrimers, i.e., amphiphilic peptides with branched structures. This, on one hand, is expected to increase antimicrobial activity of peptides owing to their interaction potential and on the other hand protect active peptides from bacterial enzymatic degradation and hence overcome resistance [92–95]. Though peptide-based dendrimers bear structural similarities with antimicrobial peptides like multiple positive charges and amphipathic nature but it is still not clear how they act at molecular level [96].

Recent efforts have been made to address these questions and new dendrimers that exhibit different potencies against microbes with respect to growth inhibition despite similar chemical structure have been designed. Dendrimer structure and antibacterial activity relation along with the ability to attack model cell membranes have been studied. These studies show that it is the supramolecular structure with respect to charge distribution and amphiphilicity and not net charge that drives membrane disruption and correlates with the dendrimers lytic activity.

For this purpose, three kinds of dendrimers were synthesized around the same core molecule and each had similar number of cationic groups. These dendrimers differed with respect to location of cationic groups (terminal versus within dendritic chain), level of branching (5 vs. 9 branches) and on the polarity of the terminal residues (Boc- vs. 2-Cl-Boc group). Dendrimers of varying chemical structure were synthesized and their antimicrobial activity against several bacterial strains was determined. Overall, three dendrimers designated as D100, D101 and

D103 were used. All dendrimers were evaluated against Gram-positive antibiotic susceptible *Staphylococcus aureus* ATCC 25923, and antibiotic resistant *Staphylococcus aureus* ATCC 43300 as well as antibiotic susceptible *Escherichia coli* ATCC 25922 and *Pseudomonas aeruginosa* ATCC 27853 Gram-negative reference strains. All dendrimers showed antimicrobial potential. It is surprising that despite same total charge these dendrimers exhibited different antimicrobial potential based on MIC values with antimicrobial potential order as D100>D101>D103. Here it is noteworthy that D101 and D103 shared similar pattern of hydrophobicity and charge with only difference being that D103 lacks chlorine atoms in the hydrophobic surface owing to which its ten times less active against both Gram-positive and Gram-negative strains as compared to D101. Hemolytic analysis showed that D100 was 2-folds less hemolytic than D101 despite exhibiting equal potential against Gram-positive bacteria.

Once the antibacterial potential of dendrimers was established along with hemolytic potential, further experiments were designed to assess the adsorption behavior of dendrimers on negatively charged silica surfaces. All dendrimers adsorbed on silica with similar affinity though D101 apparently seemed to have higher affinity that was due to binding of this dendrimer in aggregative (dimer/trimer) forms which was confirmed by rinsing with buffer, but D101 did behave differently [97–99]. After binding on silica was studied, supported lipid bilayers (SLB) were prepared by fusing vesicles of 1-palmitoyl-2-oleoyl-sn-glycero-3-phosphocholine (POPC) on silica surfaces. Binding of dendrimers to SLB was studied by exposing SLB to dendrimer solutions at various concentrations. Within one hour, steady state conditions were reached for all dendrimers except D101 at 6 μM that required >2 h. Adsorption of D100 and D103 was found to be monotonic but D101 adsorption was followed by a desorption-dominated regime after which a complex surface phenomenon where slow desorption with large changes in dissipation occurred; thereby indicating major layer rearrangement [97]. In the beginning, adsorption was independent of dendrimer but layer reorientation was specific for D101. Thus, it is reasonable to say that electrostatic forces that drive the initial dendrimer-SLB binding are same for all dendrimers owing to similar charge but concentration required for full coverage of lipid layers is different owing to the fact that saturation times were different. In binding and affinity studies, this finding that D101 behaves differently than D100 and D103 on both silica and SLBs and only this was able to induce rearrangement of the lipid bilayers was unexpected owing to the fact that D100 was the most potent antimicrobial closely followed by D101 and then by D103. The latter indeed exhibited about 10-folds higher MIC than D100 against most strains. One possible reason for D101 behavior could be the chlorine atom in its hydrophobic terminal aromatic rings coupled with its positioning of amino groups inside the molecule which makes it prone to self aggregation even after monolayers. Furthermore, major rearrangements of the SLBs by D101 is possible due to integration into the lipid bilayer that leads to membrane swelling and/or partial solubilization similar to surfactant-lipid interactions [100]. It is therefore clear that supramolecular properties of dendrimer-peptides are crucial for driving the interaction with biological membranes and antimicrobial activity in a way similar to those of antimicrobial peptides [86]. But dendrimers owing to branched structure have additional advantages such as higher stability against proteolysis, increased solubility, decreased toxicity to human cells and lower MIC [101, 102]. The correlation between dendrimer antimicrobial activities with membrane disruption potential open ways for new possible mechanisms pertaining to the net charge and the fluidity of the lipid bilayers because it has been shown that dendrimer adsorption can bring membrane permeability changes without membrane disruption and kill the cell [103].

11.7 NOW AND NEXT?

The above discussion is by no means a comprehensive account of emerging themes in dendrimer research. Almost everyday, some novel approach with clinical potential is being published (Table 11.2). Enough optimism is likely to continue in this ever evolving field so that the next two decades should witness clinically relevant diagnostic/therapeutic systems based on dendrimers.

KEYWORDS

- **affinity enhancement**
- **dendribodies**
- **dendrimer scaffolds**
- **dentistry**
- **drug resistance**
- **hydroxyapatite**

REFERENCES

1. Caminade, A. M., Turrin, C. O., Laurent, R., Ouali, A., & Delavaux-Nicot, B. (2011). *Dendrimers: Towards Catalytic, Material and Biomedical Uses*. John Wiley & Sons.
2. Page, D., Zanini, D., & Roy, R. (1996). Macromolecular recognition: Effect of multivalency in the inhibition of binding of yeast mannan to conavalin A and pea lectins by mannosylated dendrimers. *Bioorg. Med. Chem. 4*, 1949–1961.
3. Pavan, G. M., Danani, A., Pricl, S., & Smith, D. K. (2009). Modeling the multivalent recognition between dendritic molecules and DNA: understanding how ligand 'sacrifice' and screening can enhance binding. *J. Am. Chem Soc. 131*, 9686–9694.
4. Pavan, G. M. (2014). Modeling the interaction between dendrimers and nucleic acids—a molecular perspective through hierarchical scales. Chem Med Chem. 9, 2623–2631.
5. Darbre, T., & Reymond, J. L. (2006). Peptide dendrimers as artificial enzymes, receptors, and drug-delivery agents. *Acc. Chem. Res. 39*, 925–934.
6. Lee, C. C., Mackay, J. A., Frechet, J. M. J., & Szoka, F. C. (2005). Designing dendrimers for biological applications. *Nat. Biotechnol. 23*, 1517–1526.
7. Tomalia, D. A., et al. (1985). A new class of polymers—Starburst dendritic macromolecules. *Polym. J. 17*, 117–132.
8. De Brabander-Van den Berg, E. M. M., & Meijer, E. W. (1993). Poly(Propylene Imine) dendrimers—large-scale synthesis by heterogeneously catalyzed hydrogenations. *Angew. Chem. Int. Ed. 32,* 1308–1311.
9. Juttukonda, V., et al. (2006). Facile synthesis of tin oxide nanoparticles stabilized by dendritic polymers. *J. Am. Chem. Soc. 128*, 420–421.
10. Tomalia, D. A., Hall, M., & Hedstrand, D. M. (1987). Starburst dendrimers. The importance of branch junction symmetry in the development of topological shell molecules. *J. Am. Chem. Soc. 109*, 1601–1603.
11. Liao, X. L., Stellacci, F., & McGrath, D. V. (2004). Photo-switchable flexible and shape-persistent dendrimers: comparison of the interplay between a photo-chromic azobenzene core and dendrimer structure. *J. Am. Chem. Soc. 126*, 2181–2185.

12. Merkel, O., M. et al. (2009). Triazine dendrimers as non-viral gene delivery systems: effects of molecular structure on biological activity. *Bioconjugate Chem. 20,* 1799–1806.
13. Griffe, L. et al. (2007). Multiplication of human natural killer cells by nano-sized phosphonate-capped dendrimers. *Angew. Chem. Int. Ed. 46*, 2523–2526.
14. Portevin, D., et al. (2009). Regulatory activity of azabisphosphonate-capped dendrimers on human CD4-T cell proliferation for ex-vivo expansion of NK cells from PBMCs and immunotherapy. *J. Transl. Med. 7,* 82.
15. Poupot, M., et al. (2006). Design of phosphorylated dendritic architectures to promote human monocyte activation. *FASEB J. 20,* 2339–2351.
16. Fruchon, S., et al. (2009). Anti-inflammatory and immuno-suppressive activation of human monocytes by a bio-active dendrimer. *J Leukocyte Biol. 85*, 553–562.
17. Hayder, M., et al. (2011). Phosphorus-based dendrimer as nano-therapeutics targeting both inflammation and osteoclastogenesis in experimental arthritis. *Sci. Transl. Med. 3,* 81ra35.
18. Rolland, O., et al. (2009). Efficient synthesis of phosphorus-containing dendrimers capped with isosteric functions of amino-bis (methylene) phosphonic acids. *Tetrahedron Lett. 50,* 2078–2082.
19. Rolland, O., et al. (2008). Tailored control and optimization of the number of phosphonic acid termini on phosphorus-containing dendrimers for the ex-vivo activation of human monocytes. *Chem. Eur. J. 14,* 4836–4850.
20. Wang, Y., Guo, R., Cao, X., Shen, M., & Shi, X. (2011). Encapsulation of 2-methoxyestradiol within multifunctional poly(amidoamine) dendrimers for targeted cancer therapy. *Biomaterials 32*, 3322–3329.
21. Garber, S. B., Kingsbury, J. S., Gray, B. L., & Hoveyda, A. H. Efficient and recyclable monomeric and dendritic Ru-based metathesis catalysts. *J. Am. Chem. Soc. 122*, 8168–8179 (2000).
22. Nanaware-Kharade, N., Gonzalez, G. A. III, & Lay, J. O., Jr. Hendrickson Howard, P., & Peterson, E. C. (2012). Therapeutic Anti-Methamphetamine Antibody Fragment-Nanoparticle Conjugates: Synthesis and in Vitro Characterization. *Bioconjug Chem. 23,* 1864–1872.
23. Holliger, P., & Hudson, P. (2005). Engineered antibody fragments and the rise of single domains. *Nat Biotechnol 23*, 1126–1136.
24. Constantinou, A., Chen, C., & Deonarain, M. P. (2010). Modulating the pharmacokinetics of therapeutic antibodies. *Biotechnol Lett. 32*, 609–622.
25. Dolezal, O., et al. (2000). ScFv multimers of the anti-neuraminidase antibody NC10: shortening of the linker in single-chain Fv fragment assembled in V(L) to V(H) orientation drives the formation of dimers, trimers, tetramers and higher molecular mass multimers. *Protein Eng 13*, 565–574.
26. Yang, K., et al. (2003). Tailoring structure-function and pharmacokinetic properties of single-chain Fv proteins by site-specific by PEGylation. *Protein Eng 16*, 761–770.
27. Kubetzko, S., Sarkar, C. A. & Plückthun, A. (2005). Protein PEGylation decreases observed target association rates via a dual blocking mechanism. *Mol Pharmacol 68,* 1439–1454.
28. Kubetzko, S., Balic, E., Waibel, R., Zangemeister-Wittke, U. & Pluckthun, A. (2006). PEGylation and Multimerization of the Anti- p185HER-2 Single Chain Fv Fragment 4D5: Effects on: Tumor Targeting. *J Biol Chem 281,* 35186–35201.

29. Ahmad, K. M., Xiao, Y. & Soh, H. T. (2012). Selection is more intelligent than design: improving the affinity of a bivalent ligand through directed evolution. *Nucleic Acids Res 40,* 11777–11783.
30. Nanaware-Kharade, N. et al. (2015). A Nanotechnology-Based Platform for Extending the Pharmacokinetic and Binding Properties of Anti-methamphetamine Antibody Fragments. *Sci. Rep. 5*, 12060.
31. Arano, Y. (1998). Strategies to reduce renal radioactivity levels of antibody fragments. *Q J Nucl Med 42*, 262–270.
32. Jemal, A., et al. (2011). Global cancer statistics. *CA Cancer J Clin 61*, 69–90.
33. Klein, C. A. (2008). Cancer. The metastasis Cascade. *Science 321*, 1785–1787.
34. Steeg, P. S. (2006). Tumor metastasis: Mechanistic insights and clinical challenges. *Nat Med 12*, 895–904.
35. Cristofanilli, M., et al. (2004). Circulation tumor cells, disease progression and survival in metastatic breast cancer. *N Engl J Med 351,* 781–791.
36. Pantel, K., Brakenhoff, R. H., & Brandt, B. (2008). Detection, clinical relevance and specific biological properties of disseminating tumor cells. *Nat Rev Cancer 8,* 329–340.
37. Nguyen, D. X., & Bos, P. D., and Massague, J. (2009). Metastasis: From dissemination to organ specific colonization. *Nat Rev Cancer 9*, 274–284.
38. Marrinucci, D., et al. (2009). Circulating tumor cells from well differentiated lung adeno-carcinoma retain cytomorphologic features of primary tumor type. *Arch Pathol Lab Med 133*, 1468–1471.
39. Pantel, K., & Alix-Panabieres, C. (2010). Circulating tumor cells in cancer patients: Challenges and perspectives. *Trends Mol Med 16,* 398–406.
40. Zieglschmid, V., Hollmann, C., & Bocher, O. (2005). Detection of disseminated tumor cells in peripheral blood. *Crit Rev Clin Lab Sci 42*, 155–196.
41. Chiang, A. C., & Massague, J. (2008). Molecular basis of metastasis. *N Engl J Med 359,* 2814–2823.
42. Pantel, K., Alix-Panabiers, C., & Riethdorf, S. (2009). Cancer micrometastases. *Nat Rev Clin Oncol 6,* 339–351.
43. Yu, M., Stott, S., Toner, M., Maheswaran, S., & Haber, D. A. (2011). Circulating tumor cells: Approaches to isolation and characterization. *J Cell Biol 192*, 373–382.
44. Budd, G. T. et al. (2006). Circulating tumor cells versus imaging–predicting overall survival in metastatic breast cancer. *Clin Cancer Res 12*, 6403–6409.
45. Allard, W. J. et al. (2004). Tumor cells circulate in the peripheral blood of all major carcinomas but not in healthy subjects or patients with nonmalignant diseases. *Clin Cancer Res 10*, 6897–6904.
46. Chaffer, C. L. & Weinberg, R. A. (2011). A prospective on cancer cell metastasis. *A Science 331*, 1559–1564.
47. Balic, M., Williams, A., Lin, H., Datar, R. & Cote, R. J. (2013). Circulating tumor cells: From bench to bedside. *Annu Rev Med 64,* 31–44.
48. Riethdorf, S. et al. (2007). Detection of circulating tumor cells in peripheral blood of patients with metastatic breast cancer: A validation study of the cell search system. *Clin. Cancer Res 13*, 920–928.
49. Adams, A. A. et al. (2008). Highly efficient circulating tumor cell isolation from whole blood and label-free enumeration using polymer-based microfluidics with an integrated conductivity sensor. *J Am Chem Soc 130*, 8633–8641.

50. Nagrath, S., et al. (2007). Isolation of rare circulating tumor cells in cancer patients by microchip technology. *Nature 450*, 1235–1239.
51. Gleghorn, J. P. et al. (2010). Capture of circulating tumor cells from whole blood of prostate cancer patients using geometrically enhanced differential immune-capture (gedi) and a prostate-specific antibody. *Lab Chip 10,* 27–29.
52. Vona, G., et al. (2000). Isolation by size of epithelial tumor cells: A new method for the immune-morphological and molecular characterization of circulating tumor cells mol. *Am J Pathol 156*, 57–63.
53. Kuo, J. S. et al. (2010). Deformability considerations in filtration of biological cells. *Lab Chip 10,* 837–842.
54. Mohamed, H., Murray, M., Turner, J. N., & Caggana, M. (2009). Isolation of tumor cells using size and deformation. *J Chromatogr A 1216,* 8289–8295.
55. Marrinucci, D. et al. (2012). Fluid biopsy in patients with metastatic prostate, pancreatic and breast cancers. *Phys Biol 9,* 016003.
56. Dharmasiri, U. et al. (2011). High-throughput selection, enumeration, electrokinetic manipulation, and molecular profiling of low-abundance circulating tumor cells using a microfluidic system. *Anal Chem 83,* 2301–2309.
57. Sttott, S. L., et al. (2010). Isolation of Circulating tumor cells using microvortex generating herringbone-chip. *Proc Natl Acad Sci USA 107*, 18392–18397.
58. Ozkumur, E. et al. (2013). Inertial focusing for tumor antigen-dependent and independent sorting of rare circulating tumor cells, *Sci Transl. Med 5,* 179ral47.
59. Schiro, P. G., et al. (2012). Sensitive and high-throughput isolation of rare cells from peripheral blood with ensemble- decision aliquot ranking. *Angew Chem Int Ed Engl 51*, 4618–4622.
60. Myung, J. H., Gajjar, K. A., Saric, J., Eddington, D. T., & Hong, S. (2011). Dendrimer mediated multivalent binding for the enhanced capture of tumor cells. *Angew Chem Int Ed Engl 50,* 11769–11772.
61. Hong, S. et al. (2007). The binding avidity of a nanoparticle-based multivalent targeted drug delivery platform. *Chem Biol 14,* 107–115.
62. Wang, S. et al. (2009). Three-dimensional nanostructured substrates toward efficient capture of circulating tumor cells. *Angew Chem Int Ed Engl 48,* 8970–8973.
63. Klein, E., Ferrand, N. P., Barwell, A., & Davis, P. (2008). Solvent effects in carbohydrate binding by synthetic receptors: implications for the role of water in natural carbohydrate recognition. *Angew Chem Int Ed Engl. 47,* 2693–2696.
64. Cardona, C. M., & Gawley, R. E. (2002). An improved synthesis of a trifurcated newkome-type monomer and orthogonally protected two-generation dendrons. *J. Org. Chem., 67,* 1411–1413.
65. Canadian Advisory Board on Dentin Hypersensitivity. (2003). Consensus-based recommendations for the diagnosis and management of dentin hypersensitivity. *J Can Dent Assoc, 69, 221*–226.
66. Orchardson, R., & Gillam, D. G. (2006). Managing dentin hypersensitivity. *J Am Dent Assoc. 137,* 990–998.
67. Brannstrom, M., Linden, L. A., & Johnson, G. (1968). Movement of dentinal and pulpal fluid caused by clinical procedures. *J Dent Res. 47*, 679–682.
68. Markowitz, K., Pashley, D. H. (2007). Personal reflections on a sensitive subject. *J Dent Res., 86,* 292–295.

69. Brannstrom, M. (1966). Sensitivity of dentine. *Oral Surg Oral Med Oral Pathol., 21,* 517–526.
70. Morris, M. F., Davis, R. D., Richardson, B. W. (1999). Clinical efficacy of two dentin desensitizing agents. *Am J Dent. 12,* 72–76.
71. Davies, M., Paice, E. M., Jones, S. B., Leary, S., Curtis, A. R., & West, N. X. (2011). Efficacy of desensitizing dentifrices to occlude dentinal tubules. *European Eur J Oral Sci., 119,* 597–503.
72. Pillon, F. L., Romani, I. G., & Schmidt, E. R. (2004). Effect of a 3% potassium oxalate topical application on dentinal hypersensitivity after subgingival scaling and root planning. *J Periodontol. 75,* 1461–1464.
73. Gillam, D. G., Newman, H. N., Davies, E. H., Bulman, J. S., Troullos, E. S., & Curro, F. A. (2004). Clinical evaluation of ferric oxalate in relieving dentine hypersensitivity. *J Oral Rehabil. 31,* 245–250.
74. Lynch, E., Brauer, D. S., Karpukhina, N., Gillam, D. G., & Hill, R. G. (2012). Multicomponent bioactive glasses of varying fluoride content for treating dentin hypersensitivity. *Dent Mater. 28,* 168–178.
75. Suge, T., Kawasaki, A., Ishikawa, K., Matsuo, T., & Ebisu, S. (2008). Ammonium hexafluorosilicate elicits calcium phosphate precipitation and shows continuous dentin tubule occlusion. *Dent Mater. 24,* 192–198.
76. Suge, T., Ishikawa, K., Kawasaki, A., Yoshiyama, M., Asaoka, K., & Ebisu, S. (1995). Duration of dentinal tubule occlusion formed by calcium phosphate precipitation method: In vitro evaluation using synthetic saliva. *J Dent Res. 74,* 1709–1714.
77. Zhou, Z. H., Zhou, P. L., Yang, S. P., Yu, X. B., & Yang, L. Z. (2007). Controllable synthesis of hydroxyapatite nanocrystals via a dendrimer-assisted hydrothermal process. *Mater Res Bull. 42,* 1611–1618.
78. Yan, S. J., Zhou, Z. H., Zhang, F., Yang, S. P., Yang, L. Z., & Xu, X. B. (2006). Effect of anionic PAMAM with amido groups starburst dendrimers on the crystallization of $Ca_{10}(PO_4)_6(OH)_2$ by hydrothermal method. *Mater Chem Phys. 99,* 164–169.
79. Esfand, R., & Tomalia, D. A. (2001). Poly (amidoamine) (PAMAM) dendrimers: from biomimicry to drug delivery and biomedical applications. *Drug Discov Today. 6,* 427–436.
80. Tomalia, D. A. (2005). Birth of a new macromolecular architecture: dendrimers as quantized building blocks for nanoscale synthetic polymer chemistry. *Prog Polym Sci. 30,* 394–324.
81. Wang, Z. J., Sa, Y., Sauro, S., Chen, H., Xing, W. Z., Ma, X., et al. (2010). Effect of desensitizing toothpastes on dentinal tubule occlusion: A dentine permeability measurement and SEM in vitro study. *J Dent. 38,* 400–410.
82. Tanahashi, M., & Matsuda, T. (1997). Surface functional group dependence on apatite formation on self-assembled monolayers in a simulated body fluid. *J Biomed Mater Res. 34,* 305–315.
83. Toworfe, G. K., Composto, R. J., Shapiro, I. M., & Ducheyne, P. (2006). Nucleation and growth of calcium phosphate on amine-, carboxyl- and hydroxyl-silane self-assembled monolayers. *Biomaterials. 27,* 631–642.
84. Svenson, S., & Tomalia, D. A. (2012). Dendrimers in biomedical applications-reflections on the field. *Adv Drug Deliv Rev. 64,* 102–115.
85. Duncan, R., & Izzo, L. (2005). Dendrimer biocompatibility and toxicity. *Adv Drug Deliv Rev. 57* 2215–2237.

86. Schmidtchen, A., Pasupuleti, M., & Malmsten, M. (2014). Effect of hydrophobic modifications in antimicrobial peptides. *Adv. Colloid Interface Sci. 205*, 265–274.
87. Malmsten, M. (2014). Antimicrobial peptides. *Upsala, J., Med. Sci. 119,* 199–204.
88. Brogden, K. A. (2005). Antimicrobial peptides: Pore formers or metabolic inhibitors in bacteria? *Nat. Rev. Microbiol. 3,* 238–250.
89. Pasupuleti, M., Schmidtchen, A., & Malmsten, M. (2012). Antimicrobial peptides: Key components of the innate immune system. *Crit. Rev. Biotechnol. 32*, 143–171.
90. Stromstedt, A. A., Ringstad, L., Schmidtchen, A., & Malmsten, M. (2010). Interaction between amphiphilic peptides and phospholipid membranes. *Curr. Opin. Colloid Interface Sci. 15*, 467–478.
91. Hancock, R. E. W., & Sahl, H. G. (2006). Antimicrobial and host-defense peptides as new anti-infective therapeutic strategies. *Nat. Biotechnol. 24,* 1551–1557.
92. Reymond, J. L., & Darbre, T. (2012). Peptide and glycopeptide dendrimer apple trees as enzyme models and for biomedical applications. *Org. Biomol. Chem. 10,* 1483–1492.
93. Sadler, K., & Tam, J. P. (2002). Peptide dendrimers: Applications and synthesis. *Rev. Mol. Biotechnol., 90,* 195–229.
94. Crespo, L., Sanclimens, G. R., Pons, M., Giralt, E., Royo, M., & Albericio, F. (2005). Peptide and amide bond-containing dendrimers. *Chem. Rev. 105,* 1663–1682.
95. Janiszewska, J., Swieton, J., Lipkowski, A. W., & Urbanczyk-Lipkowska, Z. (2003). Low molecular mass peptide dendrimers that express antimicrobial properties. *Bioorg. Med. Chem. Lett. 13,* 3711–3713.
96. Polcyn, P, Zielinska, P., Zimnicka, M, Troć, A., Kalicki, P., Solecka, J., Laskowska, A., & Urbanczyk-Lipkowska, Z. (2013). Novel antimicrobial peptide dendrimers with amphiphilic surface and their interactions with phospholipids—Insights from mass spectrometry. *Molecules, 18,* 7120–7144.
97. Lind, T. K., Zielinska, P, Wacklin, H.P, Urbanczyk-Lipkowska, Z., & Cardenas, M. (2014). Continuous flow atomic force microscopy imaging reveals fluidity and time-dependent interactions of antimicrobial dendrimer with model lipid membranes. *ACS Nano, 8*, 396–408.
98. Sauerbrey, G. (1959). Verwendung von schwingquarzen zur wagung dunner schich ten und zur mikrowagung. *Z. Phys. 155*, 206–222.
99. Holmberg, K. J. B., Kronberg, B., & Lindman, B. (2012). *Surfactants and Polymers in Aqueous Solution, 2nd ed.*, John Wiley & Sons, Ltd.: Chichester, UK, 562.
100. Cardenas, M., Schillen, K., Alfredsson, V., Duan, R. D., Nyberg, L., & Arnebrant, T. (2008). Solubilization of sphingomyelin vesicles by addition of a bile salt. *Chem. Phys. Lipids, 151,* 10–17.
101. Tam, J. P., Lu, Y. A., & Yang, J. L. (2002). Antimicrobial dendrimeric peptides. *Eur. J. Biochem. 269,* 923–932.
102. Liu, S. P., Zhou, L., Lakshminarayanan, R., & Beuerman, R. W. (2010). Multivalent antimicrobial peptides as therapeutics: Design principles and structural diversities. *Int. J. Pept. Res. Ther. 16*, 199–213.
103. Åkesson, A., Lundgaard, C. V., Ehrlich, N., Pomorski, T. G., Stamou, D., & Cárdenas, M. (2012). Induced dye leakage by pamam g6 does not imply dendrimer entry into vesicle lumen. *Soft Matter, 8*, 8972–8980.

INDEX

D

E

F

G

H

I

L

M

N

Q

For Product Safety Concerns and Information please contact our EU representative GPSR@taylorandfrancis.com Taylor & Francis Verlag GmbH, Kaufingerstraße 24, 80331 München, Germany

Batch number: 10726519

Printed by Printforce, the Netherlands